APC Proteins

ADVANCES IN EXPERIMENTAL MEDICINE AND BIOLOGY

Recent Volumes in this Series

Volume 649
MOLECULAR MECHANISMS OF SPONDYLOARATHROPATHIES
Edited by Carlos López-Larrea, and Roberto Díaz-Peña

Volume 650
V(D)J RECOMBINATION
Edited by Pierre Ferrier

Volume 651
DEVEOPLMENT AND ENGINEERING OF DOPAMINE NEURONS
Edited by R. Jeroen Pasterkamp, Marten P. Smidt, and J. Peter H. Burbach

Volume 652
INHERITED NEUROMUSCULAR DISEASES: TRANSLATION FROM PATHOMECHANISMS TO THERAPIES
Edited by Carmen Espinós, Vicente Felipo, and Francesc Palau

Volume 653
TARGET PATTERN RECOGNITION IN INNATE IMMUNITY
Edited by Uday Kishore

Volume 654
THE ISLETS OF LANGERHANS
Edited by Md. Shahidul Islam

Volume 655
PHARMACEUTICAL BIOTECHNOLOGY
Edited by Carlos A. Guzmán and Giora Z. Feuerstein

Volume 656
APC PROTEINS
Edited by Inke S. Näthke and Brooke M. McCartney

APC Proteins

Edited by

Inke S. Näthke

Cell and Developmental Biology, College of Life Sciences, University of Dundee, Dundee, Scotland, UK

Brooke M. McCartney

Department of Biological Sciences, Carnegie Mellon University, Pittsburgh, Pennsylvania, USA

Springer Science+Business Media, LLC

Landes Bioscience

Springer Science+Business Media, LLC
Landes Bioscience

Softcover reprint of the hardcover 1st edition 2009

Springer Science+Business Media, LLC, 233 Spring Street, New York, New York 10013, USA
http://www.springer.com

Please address all inquiries to the publishers:
Landes Bioscience, 1002 West Avenue, Austin, Texas 78701, USA
Phone: 512/ 637 6050; FAX: 512/ 637 6079
http://www.landesbioscience.com

The chapters in this book are available in the Madame Curie Bioscience Database.
http://www.landesbioscience.com/curie

APC Proteins, edited by Inke S. Näthke and Brooke M. McCartney. Landes Bioscience / Springer Science+Business Media, LLC dual imprint / Springer series: Advances in Experimental Medicine and Biology.

ISBN: 978-1-4614-2451-2

While the authors, editors and publisher believe that drug selection and dosage and the specifications and usage of equipment and devices, as set forth in this book, are in accord with current recommendations and practice at the time of publication, they make no warranty, expressed or implied, with respect to material described in this book. In view of the ongoing research, equipment development, changes in governmental regulations and the rapid accumulation of information relating to the biomedical sciences, the reader is urged to carefully review and evaluate the information provided herein.

APC proteins / edited by Inke S. Näthke, Brooke M. McCartney.
p. ; cm. -- (Advances in experimental medicine and biology ; v. 656)
Includes bibliographical references and index.

1. Colon (Anatomy)--Cancer. 2. Tumor suppressor proteins. I. Näthke, Inke S. (Inke Swedlow) II. McCartney, Brooke M. (Brooke Marie) III. Series: Advances in experimental medicine and biology, v. 656. 0065-2598 ;
[DNLM: 1. Adenomatous Polyposis Coli Protein. 2. Adenomatous Polyposis Coli. 3. Colorectal Neoplasms--metabolism. 4. Genes, APC. W1 AD559 v.656 2009 / WZ 200 A639 2009]
RC280.C6A63 2009
616.99'4347--dc22

2009024855

DEDICATION

This book is dedicated to all those afflicted with colorectal cancer and to the APC research community.

PREFACE

The initial identification of the *Adenomatous polyposis coli* (*Apc*) gene as the site of mutations in familial adenomatous polyposis (FAP) was described in 1992.[1,2] A causal relationship between *Apc* mutations and intestinal tract tumours was confirmed three years later with the establishment of the Min mouse model.[3] These mice are heterozygous for *Apc* and develop numerous intestinal tumours that mimic FAP. Subsequently, *Apc* has emerged as the most commonly mutated gene in colorectal cancer with reports varying between 50-80% of sporadic tumours carrying such mutations. The search for how mutations in *Apc* initiate and/or support progression of tumours in the intestinal tract has revealed that the Apc protein is a multifunctional participant in a diverse array of cellular functions. By collecting and assembling the chapters in this book, we aimed to provide an overview of the diverse functions performed by the Apc protein. As summarised in a short final chapter by Trainer, heterozygosity of *Apc* leads to a number of extracolonic manifestations that further support this emerging picture of the Apc protein as an active contributor to many different cellular functions.

The first recognised function of Apc was its role in Wnt signalling.[4,5] This function is one of the driving forces for how mutations in *Apc* ensure that cells remain proliferative. Many of the molecular details of this pathway have been discovered and are described in the first chapter by Kennell and Cadigan. The transcriptional changes that result from *Apc* mutations due to changes in the transcriptional contribution of β-catenin are exacerbated by direct functions of Apc in the nucleus, which are discussed by Neufeld. The transcriptional changes induced by mutant Apc provide the background against which other cellular functions of Apc become particularly important in the digestive tract epithelium. Cell migration contributes significantly to the normal maintenance of this tissue, and Apc is important for efficient and directed cell migration as discussed by Manneville. Contributing to these functions is the interaction of Apc with the small microtubule binding protein EB1. This interaction and its implication for Apc are described by Morrison. The interaction between EB1 and Apc may also play a role in the ability of Apc to contribute to formation of normal spindles during mitosis, as described by Caldwell and Kaplan. How Apc affects the fabric of the mitotic spindle, namely microtubules, and how this contributes to the role of Apc in mitosis is discussed by Bahmanyar, Nelson and Barth. In addition to cell migration and proliferation, cell death is a crucial component of epithelial biology in the intestinal tract. Although the molecular mechanisms for a role for Apc in this process has not been identified, there is mounting evidence that Apc contributes to the normal execution of an apopotic program as discussed by Benchabane and Ahmed. Much of the evidence presented in these chapters is based on data obtained in cultured cells. However, important insights have also been gained using mouse model systems. These are discussed by Sansom. Furthermore, Kwong and Dove provide an overview of the genetics that relate to the role and regulation of Apc mutations in cancer and also describe a new animal model for Apc-mediated colorectal disease.

Almost every chapter includes a schematic of the Apc protein. After some consideration we decided to retain these figures in all chapters because in each case slightly different emphasis is placed on the domains that are relevant for the particular subject discussed. Furthermore, we felt that each chapter should stand on its own and thus required this information.

Colorectal cancer is one of the leading causes of cancer deaths in the Western world. It is unique because of a single genetic change, which appears to be an initiating event in a majority of cases. Understanding the underlying molecular changes that produce cancer in this tissue is crucial for our ability to develop better methods to detect and treat this disease early. Because disruption of Apc is common to most of these tumours, the biology of Apc is at the core of this problem. Colorectal cancer is one of the obvious cases where the application of basic scientific research shows great promise in providing new tools to combat a human disease.

We are extremely grateful to all authors who have provided their time and energy to support this publication and provide a comprehensive analysis of our current understanding of the complex relationships and functions performed by the Apc protein.

Inke S. Näthke, PhD
Brooke M. McCartney, PhD

1. Su LK, Kinzler KW, Vogelstein B et al. Multiple intestinal neoplasia caused by a mutation in the murine homolog of the APC gene. Science 1992; 256:668.
2. Moser AR, Dove WF, Roth KA et al. The Min (multiple intestinal neoplasia) mutation: its effect on gut epithelial cell differentiation and interaction with a modifier system. J Cell Biol 1992; 116:1517.
3. Moser AR, Luongo C, Gould KA et al. ApcMin: a mouse model for intestinal and mammary tumorigenesis. Eur J Cancer 1995; 1A:1061.
4. Rubinfeld B, Souza B, Albert I et al. Association of the APC gene product with beta-catenin. Science 1993; 262:1731.
5. Su LK, Vogelstein B, Kinzler KW. Association of the APC tumor suppressor protein with catenins. Science 1993; 262:1734.

ABOUT THE EDITORS...

INKE S. NÄTHKE is the professor of Epithelial Biology at the University of Dundee. Her research interest is to determine how specific molecular changes produce changes in cells and ultimately in whole tissues during early stages of tumourigenesis. Her focus is on colorectal cancers, but the common epithelial origin of most tumours makes this work relevant to many other organs. Her work has led her to examine not only how cells move and divide, but also how they know where they are, and how they work together to build a functional tissue.

Professor Näthke received her training at San Jose State University, the University of California San Francisco, Stanford and Harvard Medical Schools.

ABOUT THE EDITORS...

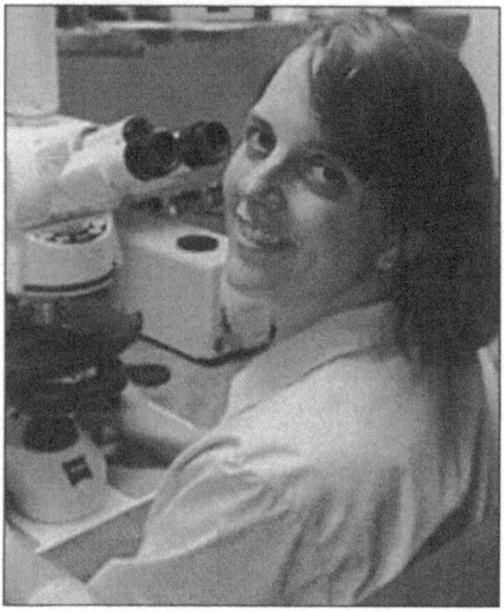

BROOKE M. MCCARTNEY is an Assistant Professor of Biological Sciences at Carnegie Mellon University. She is interested in how signaling pathways regulate the cytoskeleton and how this leads to changes in cytoskeletal and tissue architecture. The focus of her lab is the role of APC proteins and Wnt signaling in morphogenesis, both at the level of the cytoskeleton and at the level of tissues, using *Drosophila melanogaster* as a model system. Dr. McCartney received her training at Mount Holyoke College, Duke University, and the University of North Carolina at Chapel Hill.

PARTICIPANTS

Yashi Ahmed
Dartmouth Medical School
Department of Genetics
Hanover, New Hampshire
USA

Shirin Bahmanyar
Department of Biological Sciences
and
Molecular Cellular Physiology
Stanford University
Stanford, California
USA

Angela I.M. Barth
Department of Biological Sciences
and
Molecular Cellular Physiology
Stanford University
Stanford, California
USA

Hassina Benchabane
Dartmouth Medical School
Department of Genetics
Hanover, New Hampshire
USA

Kenneth M. Cadigan
Molecular, Cellular
and Developmental Biology
Natural Science Building
University of Michigan
Ann Arbor, Michigan
USA

Christine M. Caldwell
Section of Molecular
and Cellular Biology
University of California, Davis
Davis, California
USA

William F. Dove
McArdle Laboratory
for Cancer Research
and
Laboratory of Genetics
University of Wisconsin
Madison, Wisconsin
USA

Sandrine Etienne-Manneville
Cell Polarity and Migration Group
Institut Pasteur CNRS URA
Paris
France

Kenneth B. Kaplan
Section of Molecular
and Cellular Biology
University of California-Davis
Davis, California
USA

Jennifer Kennell
Molecular, Cellular
and Developmental Biology
Natural Science Building
University of Michigan
Ann Arbor, Michigan
USA

Lawrence N. Kwong
McArdle Laboratory
for Cancer Research
University of Wisconsin
Madison, Wisconsin
and
Dana-Farber Cancer Institute
Boston, Massachusetts
USA

Brooke M. McCartney
Department of Biological Sciences
Carnegie Mellon University
Pittsburgh, Pennsylvania
USA

Ewan E. Morrison
CRUK Clinical Centre at Leeds
Leeds Institute of Molecular Medicine
St James's University Hospital
Leeds, England
UK

Inke S. Näthke
Cell and Developmental Biology
College of Life Sciences
University of Dundee
Dundee, Scotland
UK

W. James Nelson
Department of Biological Sciences
and Molecular Cellular Physiology
Stanford University
Stanford, California
USA

Kristi L. Neufeld
Department of Molecular Biosciences
University of Kansas
Lawrence, Kansas
USA

Owen Sansom
Beaton Institute of Cancer Research
Glasgow, Scotland
UK

Alison H. Trainer
Prince of Wales Clinical School
University of New South Wales
Sydney
Australia

CONTENTS

Chapter 1

APC and β-Catenin Degradation

Jennifer Kennell and Kenneth M. Cadigan*

Introduction

The adenomatous polyposis coli (*APC*) gene encodes a tumor suppressor that is mutated in familial adenomatous polyposis (FAP) colon cancer as well as the large majority of sporadic colorectal cancers (reviewed in ref. 1, Chapter by Sansom, Kwong and Dove). Most of the mutations in *APC* associated with increased incidence of cancer generate a truncated form of APC.[2,3] While many studies have implicated the APC protein in the regulation of the cytoskeleton (reviewed in ref. 4, Chapter by Morrison, Caldwell and Kaplan et al.), there is also compelling evidence that APC is a key component of Wnt/β-catenin signaling (reviewed in refs. 3,5). Consistent with this, loss of *APC1* and *APC2* genes in *Drosophila* result in elevated Wnt/β-catenin signaling.[6-8] Mutations of *APC* in mice or zebrafish also increase the susceptibility of these animals to developing colorectal cancer with elevation of Wnt/β-catenin signaling in the tumors.[9,10] In this chapter, we will focus on the role of APC as a negative regulator of the Wnt/β-catenin pathway.

Wnt/β-catenin signaling is a highly conserved pathway important in many aspects of development, including axis specification and control of proliferation and differentiation (reviewed in ref. 5). In the absence of Wnt signaling, β-catenin is kept at low levels within the cytoplasm and nucleus through the coordinated action of APC and several proteins sometimes referred to as the "destruction complex". This complex causes β-catenin to be phosphorylated at specific residues, leading to ubiquitination and destruction by the proteosome. Upon Wnt stimulation, the activity of the destruction complex is inhibited, resulting in an accumulation of β-catenin in the cytosol as well as the nucleus, where it acts as a coactivator for TCF/LEF family members and other transcription factors to mediate Wnt target gene expression (reviewed in refs. 11,12). The mechanism of how Wnt signaling antagonizes the destruction complex is not fully understood, though the considerable progress made in the past decade will be presented.

APC Is a Critical Component of the β-Catenin Destruction Complex

Early studies connecting APC with β-catenin found that these two proteins physically interact.[13,14] In many colorectal cancer cell lines that contain truncated forms of APC, β-catenin is present at high levels due to increased stability. Expression of full-length APC in these cells was sufficient to downregulate β-catenin.[15] Because *APC* mutations are almost always truncations ending in a portion of the protein known as the Mutant Cluster Region (MCR; see Fig. 1), it was originally thought that the truncated proteins were necessary for β-catenin stabilization and elevated Wnt signaling or that they had dominant negative effects. However, complete loss of *APC* gene activity results in elevated β-catenin protein levels and signaling in flies[6-8] and there is no evidence for truncated APCs having dominant negative effects on Wnt signaling.[8] It has been suggested that complete removal of APC is deleterious for survival of colorectal tumors and that truncated APC

*Corresponding Author: Kenneth M. Cadigan—Associate Professor of Molecular, Cellular and Developmental Biology, Natural Science Building, University of Michigan Ann Arbor, MI. Email: cadigan@umich.edu

APC Proteins, edited by Inke S. Näthke and Brooke M. McCartney. ©2009 Landes Bioscience and Springer Science+Business Media.

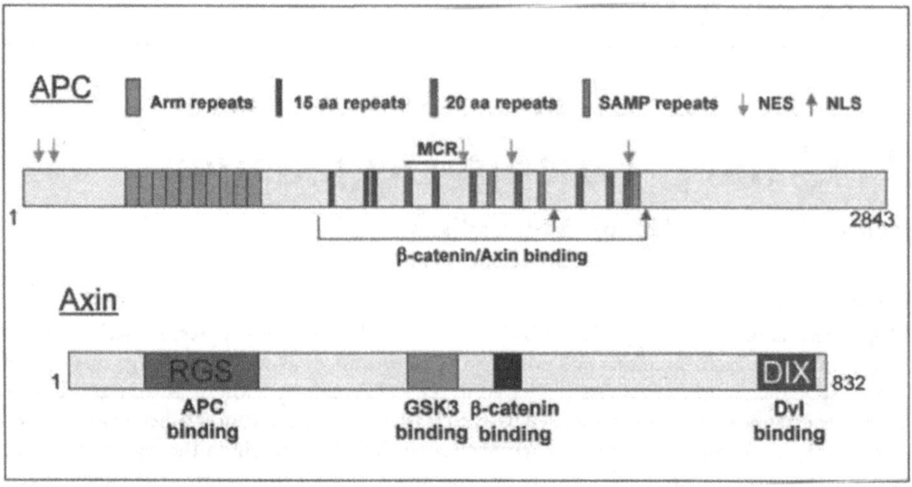

Figure 1. Diagram of the APC and Axin proteins. Cartoons of human APC and Axin are shown. Only the domains referred to in this chapter are indicated for APC. MCR refers to the Mutation Cluster Region, where a majority of APC truncations occur in colorectal cancers. Regions of human Axin that interact with other components of the destruction complex are indicated.

proteins may contain some residual activity, providing a "just right" level of elevated Wnt/β-catenin signaling.[16] However, it is also possible that the selection for APC truncations is related to its role in Wnt-independent pathways (see Chapter by Caldwell and Kaplan in this volume).

At the same time that APC was being recognized as a Wnt signaling component, additional negative regulators of β-catenin stability were identified through forward genetics. In *Drosophila*, mutations in the Glycogen Synthase Kinase 3β (GSK3β) homolog *zeste white-3/shaggy* resulted in elevated β-catenin levels and Wnt signaling.[17] In mice, characterization of the embryonic lethal mutation *fused* identified Axin as another negative regulator of the pathway acting upstream of β-catenin[18] and *Axin* mutants in flies had an identical phenotype as *zeste white-3/shaggy*.[19,20] In vitro reconstitution studies using *Xenopus* egg extracts have also demonstrated the absolute requirement for APC, Axin and GSK3β in β-catenin degradation.[21]

These regulators of β-catenin turnover have been found to physically interact through several protein-protein interaction assays. In co-immunoprecipation experiments, β-catenin, APC or GSK3β could all pull down Axin, but no evidence was found that GSK3β can bind directly to APC or β-catenin.[22-26] GSK3β directly phosphorylates and destabilizes β-catenin.[27,28] Axin can simultaneously bind to GSK3β and β-catenin and greatly facilitate GSK3β-dependent phosphorylation of β-catenin in vitro.[23,29-31] Thus Axin acts as an adaptor to bring GSK3β together with β-catenin for efficient phosphorylation.

While APC cannot directly bind to GSK3β, it does have the ability to promote modification of β-catenin by this kinase through APC binding to Axin. Axin lacking the β-catenin binding domain (AxinΔβ-catenin) cannot increase GSK3β phosphorylation of β-catenin in vitro.[26] APC alone cannot promote GSK3β-dependent phosphorylation of β-catenin, but when added with AxinΔβ-catenin a dramatic increase in β-catenin modification by GSK3β is observed.[26] The ability of Axin to interact with APC through its RGS (regulator of G protein signaling) domain (see Fig. 1) was required for β-catenin degradation.[26] These studies suggest that APC can promote efficient GSK3β phosphorylation of β-catenin by simultaneous binding to Axin and β-catenin.

While the Axin/APC scaffolds are required for efficient phosphorylation of β-catenin by GSK3β, this reaction is also greatly enhanced by prior phosphorylation of β-catenin by CKIα or CKIε.[32-34] Consistent with this, reduction of CKI activity results in elevated β-catenin levels and signaling.[33-36] A tagged form of Axin can pull down CKIα from cell extracts,[37] though it is not clear whether this interaction is direct. In addition CKIε has been shown to associate with an Axin2 (also called Conductin)/GSK3β complex though an ankyrin-repeat protein called Diversin.[38]

After phosphorylation by CKI/GSK3β, β-catenin is then ubiquitinated by β-TrCP, a subunit of the SCF ubiquitin ligase complex, which leads to degradation of β-catenin by the proteasome.[39,40] Recently, WTX, a protein encoded by a gene mutated in Wilm's tumors, was found to directly bind to both β-catenin and β-TrCP.[37] WTX could promote β-catenin degradation in Xenopus cell extracts and knockdown of WTX with siRNA caused an increase in β-catenin levels and Wnt reporter gene activity.[37] WTX also interacted with Axin and APC in co-immunoprecipation assays.[37] Taken together, these data strongly argue that WTX is another member of the destruction complex, perhaps involved in promoting β-catenin ubiquitination after it is phosphorylated.

Because the GSK3β and CKI kinases do not directly interact with β-catenin, the current view of APC and Axin is that they act as scaffolds, bringing GSK3β/CKI and β-catenin in close proximity so that phosphorylation can occur. However, the details of how these proteins functionally interact at physiological levels are not clear. Indeed, whether all of these proteins simultaneously exist as a complex within the cell remains to be demonstrated. Furthermore, while several models exist to explain the mechanism by which Wnt stimulation prevents the complex from tagging β-catenin for degradation, gaps remain, especially concerning the role of APC. This is discussed below.

The Role of APC in Recruiting β-Catenin to the Complex

The most well characterized role of APC in the destruction complex is the recruitment of β-catenin. Though Axin also binds β-catenin directly, studies suggest that both Axin and APC are required for efficient binding of β-catenin to the complex.[21] The APC protein contains a large number of domains that mediate binding to various proteins, but this chapter will focus primarily on the three types of motifs thought to be involved in binding to β-catenin or Axin. The central portion of APC consists of three 15-amino acid and seven 20-amino acid repeats (Fig. 1). These motifs are defined by a loose consensus (see ref. 41 for an alignment of both motifs). Binding of β-catenin to APC has been reported to involve both repeat regions.[42-45] Many of the reported mutations in APC result in truncation of the protein prior to the 20-amino acid repeat region (at the MCR, Fig. 1, ref. 2,3). Consistent with retaining the 15-amino acid repeats, these APC truncations can still bind to β-catenin.[15] However, they are unable to downregulate β-catenin levels, suggesting that the 20-amino acid repeat region is important for the degradation of β-catenin.[15]

While phosphorylation of β-catenin is thought to be a prerequisite for degradation, there is some evidence that the two events can be uncoupled. β-catenin is hyperphosphorylated yet stabilized in a colorectal cell line with a form of APC that is truncated prior to the third 20-amino acid repeat.[46,47] This suggests that this region including the third 20-amino acid repeat may play a role in the degradation of β-catenin but is not essential for its phosphorylation. Interestingly, although all the 15 and 20-amino acid repeats can bind to β-catenin, the third 20-amino acid repeat has the highest affinity for β-catenin in vitro.[48] It is possible that despite the presence of the other β-catenin binding domains in the APC mutant lacking the third 20-amino acid repeat, this truncation may have reduced binding for β-catenin at physiological levels of expression. Given the potential importance of this third 20-amino acid repeat in efficient β-catenin degradation, more subtle site-directed mutagenesis of this repeat in the context of full-length APC may be informative.

Three Ser-Ala-Met-Pro (SAMP) repeats in the central region of APC are required for APC binding to Axin.[22,23] Overexpression of the entire central repeat region containing the full number of 15 and 20-amino acid repeats and SAMP repeats is sufficient to bind to and downregulate β-catenin.[15] However, mice engineered with a termination codon right before the fourth 20-amino acid repeat (at position 1638; Apc1638T; only the first SAMP repeat remains) are viable and tumor free.[49] Moveover, cells from these mutants had normal β-catenin levels and Wnt reporter activity.[49,50] However, when the *APC1638T* allele is transheterozygous with *APC1572T*, a truncation removing the single SAMP repeat present in *APC1638T*, elevated nuclear β-catenin and reporter activity is observed.[49,50] These data suggest that at least one SAMP repeat is required in combination with the 15 and 20-amino acid repeats for APC to function in Wnt/β-catenin signaling.

Phosphorylation of APC Enhances Affinity for β-Catenin

Phosphorylation of APC was first reported over ten years ago.[51] Both GSK3β and CK1 can phosphorylate APC in vitro and APC phosphorylation increases its affinity for β-catenin up to 300-500 fold in vitro.[27,41,52,53] APC is phosphorylated throughout the 15 and 20-amino acid repeats.[52] However, research has focused on the third 20-amino acid repeat in particular and the role of phosphorylation of the other repeats in APC function in the destruction complex is less understood and warrants further study.

Phosphorylation of the third 20-amino acid repeat region of APC results in structuring of residues normally unstructured in the unphosphorylated state, allowing this peptide to interact with nine arm repeats of β-catenin versus five in the unphosphorylated state.[41,53] The affinity of the phosphorylated form of the third 20-amino acid repeat for β-catenin is reported to be comparable to that of E-cadherin, Lef-1 and ICAT, whereas Axin and unphosphorylated APC have much lower affinities for β-catenin.[54] The Arm repeats of β-catenin shown to bind to phosphorylated APC include the same repeats required for interaction of β-catenin with Axin. This potential overlap of binding suggests that phosphorylated APC competes with Axin for binding to β-catenin. This model will be discussed in more detail in the next section.

Given that GSK3β and CK1 already function in the destruction complex to phosphorylate β-catenin, the data cited above suggests a model where these kinases play multiple roles within the complex, first phosphorylating APC to increase its affinity for β-catenin, followed by phosphorylation of β-catenin once it is recruited. When Wnt signaling is activated, GSK3β is dissociated from the destruction complex.[55,56] The loss of GSK3β could result in a loss of APC phosphorylation, thereby decreasing it affinity for β-catenin. Phosphopeptide specific antibodies against APC, complemented by mutational analysis of known phosphorylation sites, especially those in the third 20-amino acid repeat, may provide insight into the role this modification of APC in β-catenin degradation.

A Possible Role of PP2A in Regulating Competition between APC and Axin for β-Catenin Binding

Although APC and Axin can both directly bind to β-catenin, the available evidence indicates they cannot bind simultaneously when APC is phosphorylated. The portion of β-catenin that is bound by Axin and phosphorylated APC overlap and it has been shown that Axin can compete with APC for binding to β-catenin in vitro.[41,53] These data suggest a model where the phosphorylation state of APC influences whether β-catenin is bound to APC or Axin. If the phosphorylation state of APC cycles (which is only conjecture at this point), one can imagine a system where phosphorylated APC recruits unphosphorylated β-catenin to the complex. Upon dephosphorylation of APC, β-catenin is "handed" to Axin, where it is phosphorylated by CKI/GSK3β. The phospho-β-catenin is bound by β-TrCP and ubiquitinated, while CKI/GSK3β phosphorylates APC, completing the cycle (see Fig. 2).

An attractive candidate for a phosphatase that regulates the phosphorylation state of APC is PP2A. This multimeric phosphatase has been shown to associate with Axin.[35,57,58] In addition, a regulatory subunit of PP2A, B56 (also called PR61) has been shown to bind directly to APC via the Arm repeats[59] and Axin via the amino terminus containing the RGS domain and GSK3β binding sites.[60] Overexpression of B56/PR61 (which presumably increases PP2A activity) inhibits Wnt/β-catenin signaling, though there is discrepancy as to whether this occurs through destabilization of β-catenin.[59-62] Although this controversy must be resolved, it is tempting to speculate that in the absence of PP2A, hyper-phosphorylated APC accumulates, disrupting the cycle and blocking β-catenin degradation. Consistent with this, inhibition of PP2A by okadaic acid resulted in elevated β-catenin levels in *Xenopus* embryos.[59]

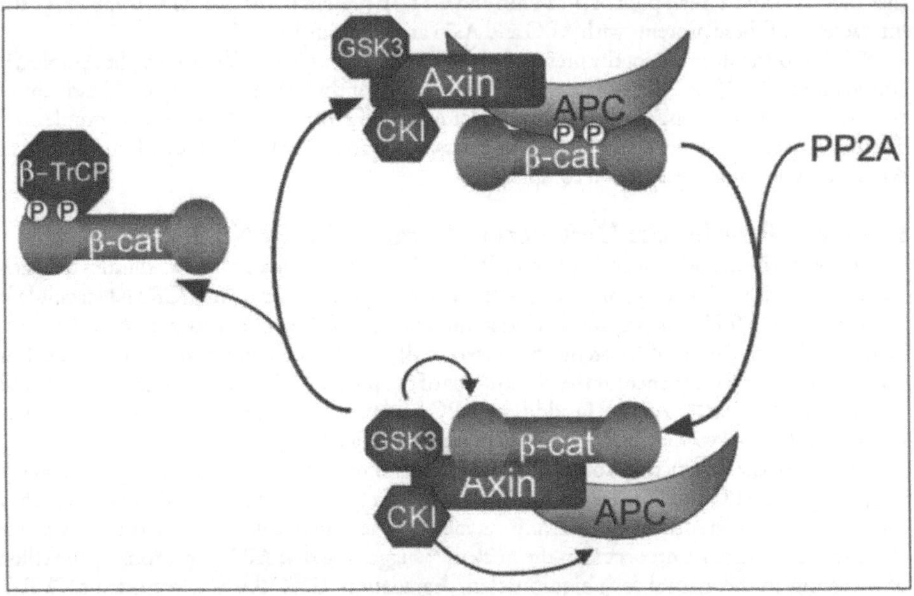

Figure 2. Speculative model for the function of the destruction complex in the absence of Wnt signaling. This model is based on the observations that APC can bind β-catenin with high affinity only when phosphorylated and that phosphorylated APC and Axin cannot bind to β-catenin simultaneously. The PP2A phosphatase complex is associated with the Axin/APC proteins and could regulate their activity by de-phosphorylating APC to allow β-catenin binding to shift from APC to Axin. GSK3β bound to Axin can phosphorylate β-catenin, allowing recognition by the F box protein β-TrCP, leading to ubiquitination and proteosomal degradation. CKI kinase is required for GSK3β-dependent phosphorylation of β-catenin and APC. Upon Wnt stimulation, this cycle is disrupted by Axin recruitment to the plasma membrane and GSK3β inhibition, which prevents β-catenin recruitment to the complex, allowing it to accumulate. See text for more details.

PP2A is likely essential for many cellular processes, making loss-of-function studies with the catalytic or structural subunits difficult. In contrast, there are numerous regulatory subunits that provide specificity for the enzyme complex.[63] Morpholino knockdown of B56ε expression in *Xenopus* embryos resulted in a decrease in Wnt/β-catenin signaling.[64] This requirement for B56ε was downstream of the Wnt ligand but upstream of Dishevelled (Dsh in flies and frogs; Dvl in mammals), a cytosolic protein required for inactivation of the destruction complex (see below). This study, combined with the work cited above,[59-62] suggests that PP2A complexes are required at multiple steps in the pathway and highlights the need for additional genetic analysis of B56 family members.

Is the Destruction Complex Really a Complex?

While this review and others assume that APC, Axin, GSK3β and β-catenin exist in a single complex, this is based on co-immunoprecipitation experiments where the proteins are overexpressed.[22,23] There have been a few reports examining whether these proteins coincide in a single complex at physiological levels and the data from these studies suggest a more complicated picture. When extracts from several cell lines were fractionated by velocity sedimentation or glycerol gradients, endogenous APC and Axin displayed significant overlap in a 23S complex, whereas GSK3β and β-catenin were predominately found in lower size fractions.[65-67] However, β-catenin can be co-immunoprecipitated with APC at endogenous levels,[67] as can GSK3β and Axin.[56] It appears that there is either a small pool of β-catenin and GSK3β present in the 23S APC complex, or the interactions of these proteins with APC and Axin are very transient.

While is difficult to detect the presence of GSK3β in any higher molecular weight complex, it appears that GSK3β activity is required for the stability of the 23S APC complex. Treatment of cells with a GSK3β inhibitors results in a shift of APC and Axin from 23S to smaller complexes.[67] These data fit with a model where GSK3β interacts, at least transiently, with the majority of cellular APC and Axin in the destruction complex.

Is There a Role for the Destruction Complex in the Nucleus?

Although the majority of endogenous APC resides in the cytoplasm,[68] APC shuttles between the cytosol and nuclear compartments, with nuclear export occurring in a CRM-1-dependent manner (ref. 69,70,71, Chapter by Neufeld in this volume). This nuclear export of APC has been linked to the regulation of β-catenin stability/activity.[72] APC contains two functional nuclear export signal (NES) sequences at the N-terminus of the protein[69,70] and three NESs in the central 20-amino acid repeats of APC.[71] In addition, APC has been reported to contain nuclear localization signals (NLSs) within the 20-amino acid repeat region.[73]

The role of the destruction complex in the nucleus is very intriguing especially given a report suggesting that APC and β-TrCP are recruited to Wnt regulated enhancers (WREs), where they somehow facilitate the dissociation of the β-catenin-Lef1 activation complex.[74] There is no evidence that β-catenin degradation occurs in the nucleus,[75] suggesting that APC may promote the efflux of β-catenin to the cytosol for ubiquitination/degradation. GSK3β is also recruited to WREs in the repressed state, but with different kinetics than APC/β-TrCP,[74] again suggesting that the components of the destruction complex are not a single entity. Like APC, Axin is also known to shuttle between nucleus and cytosol.[76,77] It will be interesting to determine whether Axin can also be detected on the chromatin of Wnt targets and whether this coincides with APC/β-TrCP or GSK3β. For a more complete discussion of the potential functions of nuclear APC please refer to Chapter by Neufeld of this volume.

Regulation of APC and the Destruction Complex by Wnt Signaling

Wnt/β-catenin signaling is transduced to the nucleus by promoting the nuclear accumulation of β-catenin. This is presumably achieved through inhibition of the destruction complex, which leads to increased stability and reduced cytoplasmic tethering of β-catenin (reviewed in ref. 78 see Fig. 3). Despite the obvious importance of the destruction complex in regulating β-catenin, the mechanism by which Wnt signaling inactivates its activity is still poorly understood.

Wnts activate the β-catenin pathway through interaction with serpentine receptors of the Frizzled (Fz) family and members of the low-density-lipoprotein-related protein (LRP) family. Several lines of evidence suggest that activation of this receptor complex generates two functionally overlapping signals, phosphorylation of Dvls and phosphorylation of the cytoplasmic tail of LRP (reviewed in ref. 11). Both of these events contribute to the inactivation of the destruction complex and have been correlated with activation of Wnt/β-catenin signaling.

Phospho-Dvl and Frat

Wnt stimulation promotes hyperphosphorylation of Dvl and this modification correlates with the ability of Dvl to activate β-catenin signaling.[79-81] But how does phosphorylation block destruction complex activity? Dvl interacts with Axin via the Dix domains of both proteins.[55,82] Dvl can also bind to cytosolic proteins called Frats,[55,83] which in turn compete with Axin for binding to GSK3β, inhibiting its ability to phosphorylate β-catenin.[31,55,84,85] Because phosphorylated Dvl binds Frat with greater affinity,[83,86] this supports a model where Wnt-mediated phosphorylation of Dvl recruits Frat to the destruction complex, where it can displace/inhibit GSK3β (see Fig. 3B).

A physiological role for Frat in Wnt/β-catenin signaling was suggested by anti-sense oligonucleotide depletion of the *Xenopus* Frat homolog GBP, which caused a loss of dorsal structures consistent with a block in the pathway.[87] However, simultaneous disruption of all three Frat

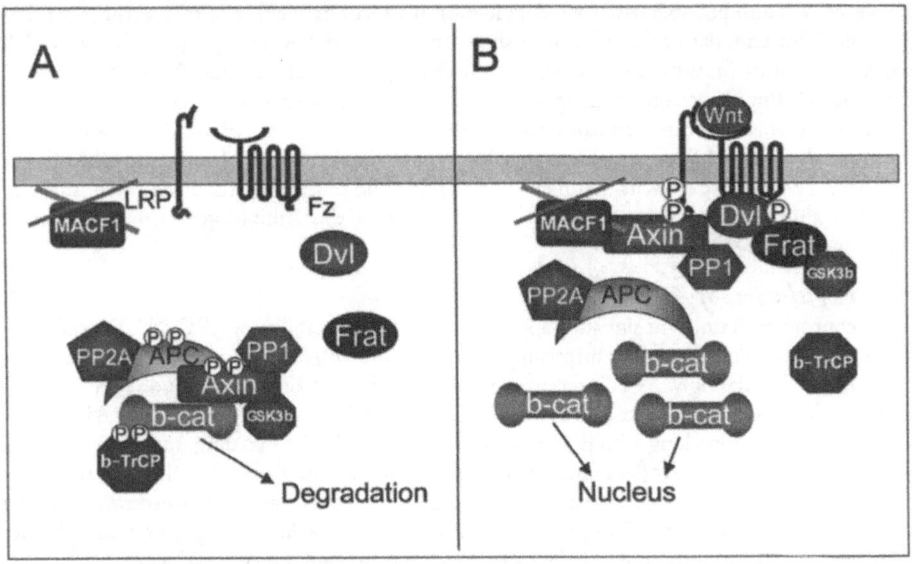

Figure 3. Model for Wnt inactivation of the destruction complex. A) In the absence of Wnt, β-catenin is bound by APC and Axin, where it is phosphorylated by CKI (not shown) and GSK3β, leading to β-TrCP mediated ubiquinitation and proteosome degradation. Phosphorylation of APC and Axin leads to greater affinity of the destruction complex for β-catenin. B) When the Wnt ligand binds to the Fz/LRP receptor complex, the cytoplasmic tail of LRP is phosphorylated by membrane associated CKI and GSK3β (not shown), leading to Axin binding. The microtubule (green line) and actin filament (red line) protein MACF1 is required for this binding. This recruitment of Axin to the plasma membrane may promote disassociation from APC. In addition, the protein phosphatases PP2A and PP1 may dephosphorylate APC and Axin, respectively. Wnt signaling also causes the membrane localization and phosphorylation of Dvl. Phospho-Dvl may remove GSK3β from the destruction complex through interaction with GSK3β binding proteins such as Frat. The end result of these events is to inhibit the ability of the complex to phosphorylate β-catenin, leading to its accumulation and nuclear translocation. See text for more details. A color version of this figure is availble on www.landesbioscience.com/curie.

genes in the mouse does not result in any obvious developmental defects, indicating normal Wnt/β-catenin signaling.[88] This dramatic result calls into question the importance of endogenous Frat in the pathway. One possibility is that other proteins exist that fulfill a similar role as Frat. One candidate is GSKIP; a GSK3β interacting protein that can bind the kinase and inhibits its ability to phosphorylate β-catenin.[89] Loss-of-function studies of GSKIP are needed to determine whether it is a *bona fide* component of the pathway.

Axin/LRP Binding

In response to Wnt binding to its coreceptors, Axin is recruited to the membrane via an interaction with LRP.[90-92] This recruitment is dependent upon CKI/GSK3β phosphorylation of the cytoplasmic tail of LRP[93,94] and requires Axin binding to Microtubule actin cross-linking factor 1 (MACF1, Fig. 3).[95] Mutagenesis of LRP revealed a tight correlation between LRP phosphorylation, binding to Axin and the ability to promote β-catenin signaling.[92-94]

In addition to binding to LRP, Axin levels are decreased upon Wnt signaling in many model systems and this decrease in stability has been reported to be GSK3β and LRP dependent.[20,90,96] These data suggest a model in which the translocation of Axin to the membrane by phosphorylated LRP helps mediate the degradation of Axin and therefore inactivation of the destruction complex.

However, it is also possible that Axin degradation is a secondary effect of this recruitment. For example, Wnt stimulation has also been shown to cause a dephosphorylation of Axin, which reduces its affinity for β-catenin.[97] Consistent with this, Protein phosphatase 1, which is required for maximal Wnt/β-catenin signaling, has been shown to dephosphorylate Axin.[98]

Is APC involved in this recruitment to the plasma membrane via LRP phosphorylation? One recent study reported Wnt-dependent translocation to the membrane of Axin and GSK3β, but not APC, in response to Wnt treatment.[95] This raises the possibility that Axin recruitment to the plasma membrane disrupts the Axin/APC interaction, thus inhibiting destruction complex function (see Fig. 3B).

Wnt Regulation of APC Stability

In contrast to Axin, Wnt signaling appears to increase the stability of APC.[99,100] The stabilization of APC by Wnt signaling is surprising given its well-established role in promoting β-catenin degradation. Intriguingly, APC is reported to be ubiquitinated in the absence of Wnt signaling and this ubiquitination is decreased upon addition of Wnts.[100] One possibility is that APC is targeted for degradation along with β-catenin in the absence of Wnt signaling, suggesting a model whereby APC "hands" over β-catenin to the degradation machinery while sacrificing itself in the process. More studies are needed to identify the role APC degradation may play in downregulating β-catenin in the absence of Wnt signaling. In addition, it will be interesting to examine whether Wnt-induced stabilization of APC plays a feedback role in regulating the pathway.

Does APC also Play a Positive Role in Wnt/β-Catenin Signaling?

Although APC is clearly a negative regulator of the Wnt/β-catenin pathway in many contexts, there are a few reports that it also has a positive effect on β-catenin signaling. Expression of APC in ventral blastomeres of *Xenopus* embryos can induce a secondary axis.[101] However, it was subsequently shown that the overexpressed APC is largely degraded, raising the possibility that truncated APC stabilizes β-catenin and induces a secondary axis by acting as a dominant negative.[84]

In *C. elegans*, the APC homolog *apr-1* gives an early embryonic phenotype similar to loss of the β-catenin homolog *wrm-1*, indicating a positive role of APC in regulating Wnt signaling.[102] Loss of *apr-1* in vulval precursor cells also gave phenotypes consistent with a reduction in Wnt/β-catenin signaling.[103] Conversely, RNAi reduction of *apr-1* was found to enhance the phenotype of *Axin* (*pry-1*) mutants, suggesting a negative role for APC in vulval induction.[104] A negative role for Apr-1 in Wnt signaling has also been found in asymmetrically dividing post-embryonic cells.[105] As in other systems, Apr-1 can physically associate with Axin/Pry-1.[106] These conflicting results suggest that APC may have both positive and negative effects on the pathway in nematodes.

Conclusion

Genetic studies in model systems ranging from invertebrates to mammals have identified APC as a negative regulator of β-catenin stability and thus Wnt/β-catenin signaling. Other negative regulators of the pathway (e.g., GSK3β and Axin) were also identified by genetics and biochemical studies have revealed that these proteins physically interact with APC to achieve β-catenin phosphorylation, leading to ubiquitination/degradation. Although it is not clear whether these proteins exist as a stable destruction complex, at the very least they interact in a dynamic way to achieve β-catenin turnover. Additional studies to more precisely identify functional domains of APC (e.g., the residues required for B56 interaction and the phosphorylation sites in the 20-amino acid repeats) will allow mechanistic models of APC function to be tested.

One limitation of studies thus far is that most functional assays for APC involve overexpression of the protein, often in colorectal cell lines containing APC truncations. While these studies have revealed that only the central portion of APC is sufficient to down-regulate β-catenin,[15] it will be informative to determine whether this is the case under physiological expression levels. For example, point mutations in the Arm repeats of fly APC2 are defective in downregulating β-catenin signaling, indicating that this portion of APC contributes to destruction complex activity.[8] Based on the model proposed in Figure 2, these mutants might be defective in binding to PP2A. More systematic mutagenesis of APC can be performed in flies, using a genomic fragment of APC2 that can rescue the APC2 mutant phenotype.[8] Additional knock-in strategies in the mouse along the lines of Smits et al (1999) will also allow APC structure-function studies to be performed in a more sophisticated way.

A better understanding of how APC and other destruction complex components promote β-catenin degradation should also provide insights into how Wnt signaling antagonizes this process. Axin recruitment to the plasma membrane through binding to LRP is clearly an important aspect of this regulation and the recent finding that APC is not recruited in a similar fashion[95] deserves further investigation. The role of Dvl remains enigmatic and the finding that *Frats* are dispensable for the pathway[88] means that additional factors remain to be identified to explain how GSK3β modification of β-catenin is inhibited. The importance of Wnt/β-catenin signaling in both normal and disease states will no doubt spur on these efforts.

References

1. Gaspar C, Fodde R. APC dosage effects in tumorigenesis and stem cell differentiation. Int J Dev Biol 2004; 48:377-86.
2. Polakis P. The oncogenic activation of beta-catenin. Curr Opin Genet Dev 1999; 9:15-21.
3. Polakis P. Wnt signaling and cancer. Genes Dev 2000; 14:1837-51.
4. Nathke IS. The adenomatous polyposis coli protein: the Achilles heel of the gut epithelium. Annu Rev Cell Dev Biol 2004; 20:337-66.
5. Logan CY, Nusse R. The Wnt signaling pathway in development and disease. Annu Rev Cell Dev Biol 2004; 20:781-810.
6. Ahmed Y, Nouri A, Wieschaus E. Drosophila Apc1 and Apc2 regulate Wingless transduction throughout development. Development 2002; 129:1751-62.
7. Akong K, Grevengoed EE, Price MH et al. Drosophila APC2 and APC1 play overlapping roles in wingless signaling in the embryo and imaginal discs. Dev Biol 2002; 250:91-100.
8. McCartney BM, Price MH, Webb RL et al. Testing hypotheses for the functions of APC family proteins using null and truncation alleles in Drosophila. Development 2006; 133:2407-18.
9. Haramis AP, Hurlstone A, van der Velden Y et al. Adenomatous polyposis coli-deficient zebrafish are susceptible to digestive tract neoplasia. EMBO Rep 2006; 7:444-9.
10. Taketo MM. Mouse models of gastrointestinal tumors. Cancer Sci 2006; 97:355-61.
11. Cadigan KM, Liu YI. Wnt signaling: complexity at the surface. J Cell Sci 2006; 119:395-402.
12. Parker DS, Blauwkamp T, Cadigan KM. Wnt/β-catenin-mediated transcriptional regulation. In: Wnt Signaling in Embryonic Development, Sokol S., ed. Advances in Developmental Biology, Wassarman PM, ed. Vol. 17, pp. 1-61. San Diego: Elsevier, 2007.
13. Rubinfeld B, Souza B, Albert I et al. Association of the APC gene product with beta-catenin. Science 1993; 262:1731-4.
14. Su LK, Vogelstein B, Kinzler KW. Association of the APC tumor suppressor protein with catenins. Science 1993; 262:1734-7.
15. Munemitsu S, Albert I, Souza B et al. Regulation of intracellular beta-catenin levels by the adenomatous polyposis coli (APC) tumor-suppressor protein. Proc Natl Acad Sci USA 1995; 92:3046-50.
16. Albuquerque C, Breukel C, van der Luijt R et al. The 'just-right' signaling model: APC somatic mutations are selected based on a specific level of activation of the beta-catenin signaling cascade. Hum Mol Genet 2002; 11:1549-60.
17. Peifer M, Sweeton D, Casey M, Wieschaus E. wingless signal and Zeste-white 3 kinase trigger opposing changes in the intracellular distribution of Armadillo. Development 1994; 120:369-80.
18. Zeng L, Fagotto F, Zhang T et al. The mouse Fused locus encodes Axin, an inhibitor of the Wnt signaling pathway that regulates embryonic axis formation. Cell 1997; 90:181-92.
19. Hamada F, Tomoyasu Y, Takatsu Y et al. Negative regulation of Wingless signaling by D-axin, a Drosophila homolog of axin. Science 1999; 283:1739-42.
20. Tolwinski NS, Wehrli M, Rives A et al. Wg/Wnt signal can be transmitted through arrow/LRP5,6 and Axin independently of Zw3/Gsk3beta activity. Dev Cell 2003; 4:407-18.
21. Salic A, Lee E, Mayer L et al. Control of beta-catenin stability: reconstitution of the cytoplasmic steps of the wnt pathway in Xenopus egg extracts. Mol Cell 2000; 5:523-32.
22. Behrens J, Jerchow BA, Wurtele M et al. Functional interaction of an axin homolog, conductin, with beta-catenin, APC and GSK3beta. Science 1998; 280:596-9.
23. Hart MJ, de los Santos R, Albert IN et al. Downregulation of beta-catenin by human Axin and its association with the APC tumor suppressor, beta-catenin and GSK3 beta. Curr Biol 1998; 8:573-81.

24. Itoh K, Krupnik VE, Sokol SY. Axis determination in Xenopus involves biochemical interactions of axin, glycogen synthase kinase 3 and beta-catenin. Curr Biol 1998; 8:591-4.
25. Nakamura T, Hamada F, Ishidate T et al. Axin, an inhibitor of the Wnt signalling pathway, interacts with beta-catenin, GSK-3beta and APC and reduces the beta-catenin level. Genes Cells 1998; 3:395-403.
26. Hinoi T, Yamamoto H, Kishida M et al. Complex formation of adenomatous polyposis coli gene product and axin facilitates glycogen synthase kinase-3 beta-dependent phosphorylation of beta-catenin and down-regulates beta-catenin. J Biol Chem 2000; 275:34399-406.
27. Rubinfeld B, Albert I, Porfiri E et al. Binding of GSK3beta to the APC-beta-catenin complex and regulation of complex assembly. Science 1996; 272:1023-6.
28. Yost C, Torres M, Miller JR et al. The axis-inducing activity, stability and subcellular distribution of beta-catenin is regulated in Xenopus embryos by glycogen synthase kinase 3. Genes Dev 1996; 10:1443-54.
29. Ikeda S, Kishida S, Yamamoto H et al. Axin, a negative regulator of the Wnt signaling pathway, forms a complex with GSK-3beta and beta-catenin and promotes GSK-3beta-dependent phosphorylation of beta-catenin. EMBO J 1998; 17:1371-84.
30. Yamamoto H, Kishida S, Uochi T et al. Axil, a member of the Axin family, interacts with both glycogen synthase kinase 3beta and beta-catenin and inhibits axis formation of Xenopus embryos. Mol Cell Biol 1998; 18:2867-75.
31. Dajani R, Fraser E, Roe SM et al. Structural basis for recruitment of glycogen synthase kinase 3beta to the axin-APC scaffold complex. EMBO J 2003; 22:494-501.
32. Amit S, Hatzubai A, Birman Y et al. Axin-mediated CKI phosphorylation of beta-catenin at Ser 45: a molecular switch for the Wnt pathway. Genes Dev 2002; 16: 1066-76.
33. Liu C, Li Y, Semenov M et al. Control of beta-catenin phosphorylation/degradation by a dual-kinase mechanism. Cell 2002; 108:837-47.
34. Yanagawa S, Matsuda Y, Lee JS et al. Casein kinase I phosphorylates the Armadillo protein and induces its degradation in Drosophila. EMBO J 2002; 21:1733-42.
35. Gao ZH, Seeling JM, Hill V et al. Casein kinase I phosphorylates and destabilizes the beta-catenin degradation complex. Proc Natl Acad Sci USA 2002; 99:1182-7.
36. Zhang L, Jia J, Wang B et al. Regulation of wingless signaling by the CKI family in Drosophila limb development. Dev Biol 2006; 299:221-37.
37. Major MB, Camp ND, Berndt JD et al. Wilms tumor suppressor WTX negatively regulates WNT/beta-catenin signaling. Science 2007; 316:1043-6.
38. Schwarz-Romond T, Asbrand C, Bakkers J et al. The ankyrin repeat protein Diversin recruits Casein kinase Iepsilon to the beta-catenin degradation complex and acts in both canonical Wnt and Wnt/JNK signaling. Genes Dev 2002; 16:2073-84.
39. Hart M, Concordet JP, Lassot I et al. The F-box protein beta-TrCP associates with phosphorylated beta-catenin and regulates its activity in the cell. Curr Biol 1999; 9:207-10.
40. Kitagawa M, Hatakeyama S, Shirane M et al. An F-box protein, FWD1, mediates ubiquitin-dependent proteolysis of beta-catenin. EMBO J 1999; 18:2401-10.
41. Ha NC, Tonozuka T, Stamos JL et al. Mechanism of phosphorylation-dependent binding of APC to beta-catenin and its role in beta-catenin degradation. Mol Cell 2004; 15:511-21.
42. Rubinfeld B, Souza B, Albert I et al. The APC protein and E-cadherin form similar but independent complexes with alpha-catenin, beta-catenin and plakoglobin. J Biol Chem 1995; 270: 5549-55.
43. Nakagawa H, Murata Y, Koyama K et al. Identification of a brain-specific APC homologue, APCL and its interaction with beta-catenin. Cancer Res 1998; 58:5176-81.
44. McCartney BM, Dierick HA, Kirkpatrick C et al. Drosophila APC2 is a cytoskeletally-associated protein that regulates wingless signaling in the embryonic epidermis. J Cell Biol 1999; 146:1303-18.
45. Spink KE, Polakis P, Weis WI. Structural basis of the Axin-adenomatous polyposis coli interaction. EMBO J 2000; 19:2270-9.
46. Sadot E, Conacci-Sorrell M, Zhurinsky J et al. Regulation of S33/S37 phosphorylated beta-catenin in normal and transformed cells. J Cell Sci 2002; 115:2771-80.
47. Yang J, Zhang W, Evans PM et al. Adenomatous polyposis coli (APC) differentially regulates beta-catenin phosphorylation and ubiquitination in colon cancer cells. J Biol Chem 2006; 281:17751-7.
48. Liu J, Xing Y, Hinds TR et al. The third 20 amino acid repeat is the tightest binding site of APC for beta-catenin. J Mol Biol 2006; 360:133-44.
49. Smits R, Kielman MF, Breukel C et al. Apc1638T: a mouse model delineating critical domains of the adenomatous polyposis coli protein involved in tumorigenesis and development. Genes Dev 1999; 13:1309-21.
50. Kielman MF, Rindapaa M, Gaspar C et al. Apc modulates embryonic stem-cell differentiation by controlling the dosage of beta-catenin signaling. Nat Genet 2002; 32:594-605.

51. Bhattacharya G, Boman BM. Phosphorylation of the adenomatous polyposis coli protein and its possible regulatory effects in cells. Biochem Biophys Res Commun 1995; 208:103-10.
52. Rubinfeld B, Tice DA, Polakis P. Axin-dependent phosphorylation of the adenomatous polyposis coli protein mediated by casein kinase 1epsilon. J Biol Chem 2001; 276:39037-45.
53. Xing Y, Clements WK, Le Trong I et al. Crystal structure of a beta-catenin/APC complex reveals a critical role for APC phosphorylation in APC function. Mol Cell 2004; 15:523-33.
54. Choi HJ, Huber AH, Weis WI. Thermodynamics of beta-catenin-ligand interactions: the roles of the N- and C-terminal tails in modulating binding affinity. J Biol Chem 2006; 281:1027-38.
55. Li L, Yuan H, Weaver CD et al. Axin and Frat1 interact with dvl and GSK, bridging Dvl to GSK in Wnt-mediated regulation of LEF-1. EMBO J 1999; 18:4233-40.
56. Liu X, Rubin JS, Kimmel AR. Rapid, Wnt-induced changes in GSK3beta associations that regulate beta-catenin stabilization are mediated by Galpha proteins. Curr Biol 2005; 15:1989-97.
57. Hsu W, Zeng L, Costantini F. Identification of a domain of Axin that binds to the serine/threonine protein phosphatase 2A and a self-binding domain. J Biol Chem 1999; 274:3439-45.
58. Ikeda S, Kishida M, Matsuura Y et al. GSK-3beta-dependent phosphorylation of adenomatous polyposis coli gene product can be modulated by beta-catenin and protein phosphatase 2A complexed with Axin. Oncogene 2000; 19:537-45.
59. Seeling JM, Miller JR, Gil R et al. Regulation of beta-catenin signaling by the B56 subunit of protein phosphatase 2A. Science 1999; 283:2089-91.
60. Yamamoto H, Hinoi T, Michiue T et al. Inhibition of the Wnt signaling pathway by the PR61 subunit of protein phosphatase 2A. J Biol Chem 2001; 276:26875-82.
61. Ratcliffe MJ, Itoh K, Sokol SY. A positive role for the PP2A catalytic subunit in Wnt signal transduction. J Biol Chem 2000; 275:35680-3.
62. Li X, Yost HJ, Virshup DM et al. Protein phosphatase 2A and its B56 regulatory subunit inhibit Wnt signaling in Xenopus. EMBO J 2001; 20:4122-31.
63. McCright B, Virshup DM. Identification of a new family of protein phosphatase 2A regulatory subunits. J Biol Chem 1995; 270:26123-8.
64. Yang J, Wu J, Tan C et al. PP2A:B56epsilon is required for Wnt/beta-catenin signaling during embryonic development. Development 2003; 130:5569-78.
65. Kishida M, Koyama S, Kishida S et al. Axin prevents Wnt-3a-induced accumulation of beta-catenin. Oncogene 1999a; 18:979-85.
66. Reinacher-Schick A, Gumbiner BM. Apical membrane localization of the adenomatous polyposis coli tumor suppressor protein and subcellular distribution of the beta-catenin destruction complex in polarized epithelial cells. J Cell Biol 2001; 152:491-502.
67. Penman GA, Leung L, Nathke IS. The adenomatous polyposis coli protein (APC) exists in two distinct soluble complexes with different functions. J Cell Sci 2005; 118:4741-50.
68. Brocardo M. Nathke IS, Henderson BR. Redefining the subcellular location and transport of APC: new insights using a panel of antibodies. EMBO Rep 2005; 6:184-90.
69. Henderson BR. Nuclear-cytoplasmic shuttling of APC regulates beta-catenin subcellular localization and turnover. Nat Cell Biol 2000; 2:653-60.
70. Neufeld KL, Nix DA, Bogerd H et al. Adenomatous polyposis coli protein contains two nuclear export signals and shuttles between the nucleus and cytoplasm. Proc Natl Acad Sci USA 2000a; 97:12085-90.
71. Rosin-Arbesfeld R, Townsley F, Bienz M. The APC tumour suppressor has a nuclear export function. Nature 2000; 406:1009-12.
72. Rosin-Arbesfeld R, Cliffe A, Brabletz T et al. Nuclear export of the APC tumour suppressor controls beta-catenin function in transcription. EMBO J 2003; 22:1101-13.
73. Zhang F, White RL, Neufeld KL. Phosphorylation near nuclear localization signal regulates nuclear import of adenomatous polyposis coli protein. Proc Natl Acad Sci USA 2000; 97:12577-82.
74. Sierra J, Yoshida T, Joazeiro CA et al. The APC tumor suppressor counteracts beta-catenin activation and H3K4 methylation at Wnt target genes. Genes Dev 2006; 20:586-600.
75. Neufeld KL, Zhang F, Cullen BR et al. APC-mediated downregulation of beta-catenin activity involves nuclear sequestration and nuclear export. EMBO Rep 2000b; 1:519-23.
76. Cong F, Varmus H. Nuclear-cytoplasmic shuttling of Axin regulates subcellular localization of beta-catenin. Proc Natl Acad Sci USA 2004; 101:2882-7.
77. Wiechens N, Heinle K, Englmeier L et al. Nucleo-cytoplasmic shuttling of Axin, a negative regulator of the Wnt-beta-catenin Pathway. J Biol Chem 2004; 279:5263-7.
78. Tolwinski NS, Wieschaus E. Rethinking WNT signaling. Trends Genet 2004; 20:177-81.
79. Yanagawa S, van Leeuwen F, Wodarz A et al. The disheveled protein is modified by Wingless signaling in Drosophila. Genes Dev 1995; 9:1087-97.

80. Lee JS, Ishimoto A, Yanagawa S. Characterization of mouse dishevelled (Dvl) proteins in Wnt/Wingless signaling pathway. J Biol Chem 1999; 274:21464-70.
81. Gonzalez-Sancho JM, Brennan KR, Castelo-Soccio LA et al. Wnt proteins induce dishevelled phosphorylation via an LRP5/6- independent mechanism, irrespective of their ability to stabilize beta-catenin. Mol Cell Biol 2004; 24:4757-68.
82. Kishida S, Yamamoto H, Hino S et al. DIX domains of Dvl and axin are necessary for protein interactions and their ability to regulate beta-catenin stability. Mol Cell Biol 1999b; 19:4414-22.
83. Hino S, Michiue T, Asashima M et al. Casein kinase I epsilon enhances the binding of Dvl-1 to Frat-1 and is essential for Wnt-3a-induced accumulation of beta-catenin. J Biol Chem 2003; 278:14066-73.
84. Farr GH 3rd, Ferkey DM, Yost C et al. Interaction among GSK-3, GBP, axin and APC in Xenopus axis specification. J Cell Biol 2000; 148:691-702.
85. Ferkey DM, D Kimelman. Glycogen synthase kinase-3 beta mutagenesis identifies a common binding domain for GBP and Axin. J Biol Chem 2002; 277:16147-52.
86. Lee E, Salic A, Kirschner MW. Physiological regulation of [beta]-catenin stability by Tcf3 and CK1epsilon. J Cell Biol 2001; 154:983-93.
87. Yost C, Farr GH 3rd, Pierce SB et al. GBP, an inhibitor of GSK-3, is implicated in Xenopus development and oncogenesis. Cell 1998; 93:1031-41.
88. van Amerongen R, Nawijn M, Franca-Koh J et al. Frat is dispensable for canonical Wnt signaling in mammals. Genes Dev 2005; 19:425-30.
89. Chou HY, Howng SL, Cheng TS et al. GSKIP is homologous to the Axin GSK3beta interaction domain and functions as a negative regulator of GSK3beta. Biochemistry 2006; 45:11379-89.
90. Mao J, Wang J, Liu B et al. Low-density lipoprotein receptor-related protein-5 binds to Axin and regulates the canonical Wnt signaling pathway. Mol Cell 2001; 7:801-9.
91. Cliffe A, Hamada F, Bienz M. A role of Dishevelled in relocating Axin to the plasma membrane during wingless signaling. Curr Biol 2003; 13:960-6.
92. Tamai K, Zeng X, Liu C et al. A mechanism for Wnt coreceptor activation. Mol Cell 2004; 13:149-56.
93. Davidson G, Wu W, Shen J et al. Casein kinase 1 gamma couples Wnt receptor activation to cytoplasmic signal transduction. Nature 2005; 438:867-72.
94. Zeng X, Tamai K, Doble B et al. A dual-kinase mechanism for Wnt coreceptor phosphorylation and activation. Nature 2005; 438:873-7.
95. Chen HJ, Lin CM, Lin CS et al. The role of microtubule actin cross-linking factor 1 (MACF1) in the Wnt signaling pathway. Genes Dev 2006; 20:1933-45.
96. Yamamoto H, Kishida S, Kishida M et al. Phosphorylation of axin, a Wnt signal negative regulator, by glycogen synthase kinase-3beta regulates its stability. J Biol Chem 1999; 274:10681-4.
97. Willert K, Shibamoto S, Nusse R. Wnt-induced dephosphorylation of axin releases beta-catenin from the axin complex. Genes Dev 1999; 13: 1768-73.
98. Luo W, Peterson A, Garcia BA et al. Protein phosphatase 1 regulates assembly and function of the beta-catenin degradation complex. EMBO J 2007; 26:1511-21.
99. Papkoff J, Rubinfeld B, Schryver B et al. Wnt-1 regulates free pools of catenins and stabilizes APC-catenin complexes. Mol Cell Biol 1996; 16:2128-34.
100. Choi J, Park SY, Costantini F et al. Adenomatous polyposis coli is down-regulated by the ubiquitin-proteasome pathway in a process facilitated by Axin. J Biol Chem 2004; 279:49188-98.
101. Vleminckx K, Wong E, Guger K et al. Adenomatous polyposis coli tumor suppressor protein has signaling activity in Xenopus laevis embryos resulting in the induction of an ectopic dorsoanterior axis. J Cell Biol 1997; 136:411-20.
102. Rocheleau CE, Downs WD, Lin R et al. Wnt signaling and an APC-related gene specify endoderm in early C. elegans embryos. Cell 1997; 90:707-16.
103. Hoier EF, Mohler WA, Kim SK et al. The Caenorhabditis elegans APC-related gene apr-1 is required for epithelial cell migration and Hox gene expression. Genes Dev 2000; 14:874-86.
104. Gleason JE, Korswagen HC, Eisenmann DM. Activation of Wnt signaling bypasses the requirement for RTK/Ras signaling during C. elegans vulval induction. Genes Dev 2002; 16:1281-90.
105. Mizumoto K, Sawa H. Cortical beta-catenin and APC regulate asymmetric nuclear beta-catenin localization during asymmetric cell division in C. elegans. Dev Cell 2007; 12:287-99.
106. Korswagen HC, Coudreuse DY, Betist MC et al. The Axin-like protein PRY-1 is a negative regulator of a canonical Wnt pathway in C. elegans. Genes Dev 2002; 16:1291-302.

CHAPTER 2

Nuclear APC

Kristi L. Neufeld*

Abstract

Mutational inactivation of the tumor suppressor gene *APC* (Adenomatous polyposis coli) is thought to be an initiating step in the progression of the vast majority of colorectal cancers. Attempts to understand APC function have revealed more than a dozen binding partners as well as several subcellular localizations including at cell-cell junctions, associated with microtubules at the leading edge of migrating cells, at the apical membrane, in the cytoplasm and in the nucleus. The present chapter focuses on APC localization and functions in the nucleus. APC contains two classical nuclear localization signals, with a third domain that can enhance nuclear import. Along with two sets of nuclear export signals, the nuclear localization signals enable the large APC protein to shuttle between the nucleus and cytoplasm. Nuclear APC can oppose β-catenin-mediated transcription. This down-regulation of nuclear β-catenin activity by APC most likely involves nuclear sequestration of β-catenin from the transcription complex as well as interaction of APC with transcription corepressor CtBP. Additional nuclear binding partners for APC include transcription factor activator protein AP-2α, nuclear export factor Crm1, protein tyrosine phosphatase PTP-BL and perhaps DNA itself. Interaction of APC with polymerase β and PCNA, suggests a role for APC in DNA repair. The observation that increases in the cytoplasmic distribution of APC correlate with colon cancer progression suggests that disruption of these nuclear functions of APC plays an important role in cancer progression. APC prevalence in the cytoplasm of quiescent cells points to a potential function for nuclear APC in control of cell proliferation. Clear definition of APC's nuclear function(s) will expand the possibilities for early colorectal cancer diagnostics and therapeutics targeted to APC.

Introduction

The tumor suppressor APC is a ~310 kDa protein with minimal sequence homology to other characterized proteins. Mutation of the *APC* gene is considered to be an early, if not the first step in the progression of more than 80% of all colorectal cancers, both inherited and sporadic. In an attempt to understand why APC function is so critical for suppression of colorectal cancers, several research groups used immunofluorescence microscopy to establish the subcellular localization pattern of APC in normal epithelial cells. The initial report characterizing APC in the cell's nucleus[1] was received with some skepticism, due to previous reports of microtubule-associated APC at the leading edge of cells.[2-4] Nevertheless, in the decade since the initial description of nuclear APC, research papers from more than 20 labs have confirmed the nuclear localization and have added to our understanding of the function of APC in the nucleus.

Detection of Nuclear APC

The first characterization of nuclear APC in human epithelial cells was published six years after identification of the *APC* gene. The nuclear APC localization was confirmed using four

*Kristi L. Neufeld—Department of Molecular Biosciences, University of Kansas, Lawrence, KS, 66045 USA. Email: klneuf@ku.edu

APC Proteins, edited by Inke S. Näthke and Brooke M. McCartney.

different APC antibodies and several cell lines for immunofluorescence microscopy and cell fractionation.[1] Nuclear APC had previously been detected in the villus epithelial cells of normal mice as well as at the cell's periphery and throughout the cytoplasm.[5] Subsequently, nuclear APC was also reported in normal human colon tissue as detected by confocal immunofluorescence microscopy[6] and immunogold electron microscopy.[7] In addition to the nuclear staining, prominent cell border staining was detected in nearly all of the images gathered from more than 50 patient samples.[6] It is worth noting that in cultured cells, APC typically appears in the nucleus, in large puncta at the leading edge and in smaller puncta throughout the cytoplasm. In contrast, APC in normal human colon tissue appears in the nucleus and at the cell border, with little if any apparent cytoplasmic localization.

Initial examination of APC localization also revealed a nucleolar accumulation of APC in some cultured cells.[1] Ectopically expressed full-length APC has also been described to concentrate in the nucleoli, especially with inhibition of nuclear export.[8] Although observed using two independent methods, the nucleolar APC distribution appears rather transient and thus it remains unclear what role APC may have in the nucleoli. Regardless of the precise subnuclear localization, overwhelming data now support the idea that APC resides in the nucleus (Table 1A).[1,5-8]

APC Domains Contributing to Nuclear Import and Export

Nearly all *APC* mutations associated with colorectal cancer result in production of a truncated APC protein, typically including only the N-terminal 25-50% of APC. In contrast to full-length APC, a truncated form of APC expressed in a colon cancer cell line did not display a nuclear distribution.[1] Therefore, it was proposed that two potential nuclear localization signals (NLS) identified in the C-terminal half of APC and missing from the truncated form, were responsible for the different subcellular distributions of full-length versus truncated APC. At 2843 amino acids (~310 kDa), APC is much too large to enter the nucleus by passive diffusion which has an estimated size limitation of 30-50 kDa.[9] It is now known that APC contains two classic monopartite basic NLSs in the C-terminal half of the protein and at least one additional domain to facilitate its nuclear import (NLS-Arm, Figs. 1A,C, Table 1B).[10,11] This additional domain of APC includes four of the seven Armadillo-repeat sequences comprising what is known as the ARM domain of

Table 1A. Evidence for nuclear APC

Observation	Cell Line(s)	Antibodies	Ref
APC in nuclear puncta by immunofluorescence (IF) microscopy (also cytoplasmic and at leading edge)	184A1, T47D	Ab-2,-4,-6, APC64	1
Cell fractionation/immunoblot: APC in nuclear and nuclear matrix fractions	LS174T, HCT116	Ab-1, -2	1
truncated APC not in nuclear fractions	DLD-1	Ab-1	1
APC in nuclei of murine villus epithelial cells IF with confocal microscopy	mouse tissue	α-APC2 (middle)	5
APC in nuclei of human colonic epithelial cells IF with confocal microscopy	human tissue, 50 patient samples	10 different	6
APC in nuclei of human colonic epithelial cells Immunogold electron microscopy	human tissue	C-APC 28.9	7
Nuclear APC colocalizes with rRNA (nucleoli)	184A1	Ab-4	1
APC in nucleoli when ectopically expressed, and nuclear export inhibited by leptomycin B (LMB)	3T3, HCT116	Ab-7	8

Table 1B. Evidence for APC's nuclear import signals (NLS)

	Ref
NLS1 (aa. 1767-1772) GKKKKP	
1. Targets heterologous cytoplasmic protein β-Gal for nuclear import alanine substitutions for all 4 lysines abolished this ability	11
2. 2 NLS1 fused in tandem mediated nuclear import more efficiently than one	
3. Mutation in context of full-length APC compromises nuclear import	
NLS2 (aa. 2048-2053) PKKKKP	
1. Targets heterologous cytoplasmic protein β-Gal for nuclear import alanine substitutions for first 2 lysines abolished this ability	11
2. Mutation in context of full-length APC compromises nuclear import	
3. Activity blocked by phosphorylation of potential PKA site and enhanced by phosphorylation of flanking CDK2 site in NLS-β-Gal fusion and in full-length APC, exogenously expressed	
NLS-Arm (aa. 334-625)	
1. Endogenous truncated APC lacking NLS1 and NLS2 can localize to the nucleus	10
2. Truncated APC can localize to the nucleus when expressed exogenously, nuclear localization was enhanced with co-expression of B56α although not dependent on phosphatase activity	

APC (aa. 453-767, Fig. 1A). That the ARM domain could mediate nuclear import is not without precedence, as the β-catenin/Armadillo protein contains 12 Armadillo repeats and the three most C-terminal repeats, along with the C-terminal 116 aa tail appear to play an essential role in nuclear import of β-catenin.[12] Nuclear import of β-catenin can occur in the absence of soluble factors or energy sources and has been proposed to be mediated through direct interaction with a nuclear pore component.[13,14]

Given that APC contains at least three nuclear localization domains, one might expect APC to be exclusively nuclear. Some potential explanations for cytoplasmic and cell junctional APC include regulation of the NLS activity, retention of APC at various locations by other protein partners and the presence of intrinsic nuclear export signals. In fact, all three of these factors likely influence the ultimate subcellular localization of APC.

As with nuclear import, proteins larger than 30-50 kDa cannot export through the nuclear pore by passive diffusion. Many proteins known to actively export from the nucleus contain leucine-rich amino acid sequences that mediate interaction with the nuclear export factor Crm1. APC contains five such classical nuclear export signals (NES), grouped at the N-terminus and in the central repeat region of the protein (Fig. 1A,C). The two NESs initially identified near the N-terminus of APC were shown to function using a wide variety of export assays (summarized in Table 1C).[8,15,16]

The identification of NESs in the central domain of APC was first made in Drosophila APC2/E-APC (dAPC2, Fig. 1B). At 1067 amino acids, dAPC2 is the smaller of the two Drosophila APC family members. dAPC2 shares 57% sequence identity with human APC in the ARM domain and contains similar features in the rest of the protein including 15-amino acid repeats, 20-amino acid repeats (20R) and SAMP repeats (Fig. 1B).[17,18] Two NESs were identified in the 3rd and 4th 20R of dAPC2.[19] These are conserved in human APC which contains an additional putative NES in the 7th 20R (Figs. 1A,C). Evidence supporting the activity of these central NESs is also summarized in Table 1C.[19,20]

When all five NESs of human APC were analyzed side-by-side in a GFP-based nuclear export assay, NES1 appeared clearly dominant and it was proposed that NES1 is the primary NES in

Table 1C. Evidence for APC's nuclear export signals (NES)

	Ref
NES1 (aa. 68-77) and NES2 (aa. 165-174)	
1. Each directs nuclear export when fused to GFP-tagged HIV-Rev with inactivated endogenous NES	8,15
2. Each directs nuclear export when fused to purified GST and injected into nuclei of living cells	15
3. Each can functionally substitute for HIV-Rev NES in in vivo Rev activity assay	15
4. Together, mediated Crm1 interaction with a short N-terminal fragment of APC (aa. 1-270) in a mammalian two-hybrid assay	15
5. The two NESs were able to facilitate nuclear shuttling of APC fragment (aa 1-270) in a heterokaryon assay	15
6. Substitution of alanine for critical leucine residues in each NES inhibited the activities measured in all assays listed (1-5)	8,15
7. Mutation of either NES in the context of full-length exogenous APC resulted in shift of APC from predominantly cytoplasmic to more nuclear than cytoplasmic	8,15
8. Purified monomeric fragment of APC (aa. 129-250) containing only NES2 bound the nuclear exportin protein Crm1 in vitro	16
NES-R3 (aa. 1506-1511), NES-R4 (aa. 1657-1662) and NES-R7 (aa. 2027-2032)	
1. Each directs nuclear export when fused to GFP-tagged N-terminal fragment of Drosophila APC2, dAPC2 (aa. 114-410)	19
2. dAPC2 (aa. 460-1067) containing NES-R3 and -R4 and fused with GFP was cytoplasmic when expressed in COS cells, but more nuclear when cells were treated with LMB	19
3. APC (aa. 1379-2080) containing NES-R3, -R4 and -R7 behaved similarly	19
4. APC (aa. 1379-2080) showed a loss of 50% nuclear fluorescence after 20 seconds of sequential photobleaching of the cytoplasm	20

mediating nuclear export of APC.[10] However, both of the N-terminal NESs appear active by a wide range of criteria (see Table 1C) and appear able to mediate nuclear-cytoplasmic shuttling of truncated APC lacking the three 20R NESs. Alternatively, it has been argued that the middle 20R NESs predominantly control localization of full-length APC protein.[20] The initial appeal of this hypothesis came from the observation that the 20R-NESs are eliminated from the truncated APC expressed in most colorectal cancers.[19] This model predicts that truncated APC lacking the central NESs would be export-compromised and nuclear. Indeed, APC truncated before NES-R3 was reported to be at least partially nuclear by immunofluorescence microscopic analysis of nine

Figure 1, viewed on following page. Schematic representation of APC domains as they relate to nuclear functions. A) Human APC is 2843 amino acids, with binding sites for several proteins shown on the middle panel. The 15 aa repeats (15R) and 20 aa repeats (20R) are involved in β-catenin binding and degradation. The lower panel highlights regions of APC that facilitate nuclear-cytoplasmic shuttling. Two NESs are located near the N-terminus, with three more in the 20aa repeat domain in the center of APC. Two NLSs are located in the central domain near axin binding sites. An additional region of APC that is able to facilitate nuclear import of truncated APC is shown, NLS-Arm. B) Drosophila APC2 (dAPC2) is 1067 amino acids, with conserved domains labeled and NES-R3 and -R4 indicated (*). Figures are to scale, with amino acid increments indicated at the top. C) Nuclear localization signals and nuclear export sequences in human APC. Numbers for the initial amino acid in each sequence are shown. Boxes indicate important hydrophobic residues (leucine or isoleucine).

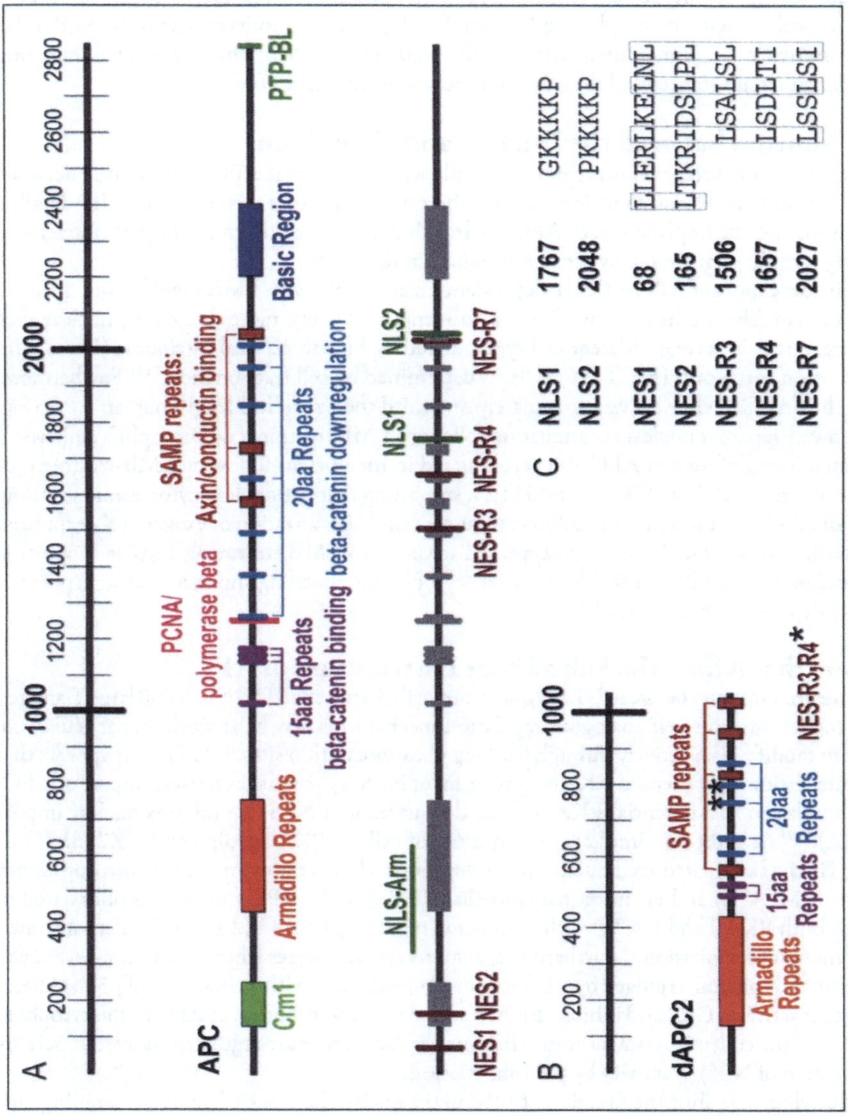

Figure 1. Legend viewed on previous page.

colorectal cancer cell lines. Consistent with this observation, cells expressing APC truncations that include NES-R3 showed nuclear exclusion of APC.[20] However, these results conflict directly with observations that endogenous truncated APC lacking NES-R3 found in SW480 colon cancer cells localizes predominantly in the cytoplasmic fraction, with a clear shift into the nuclear fraction following treatment with an inhibitor of Crm1-dependent nuclear export, leptomycin B (LMB).[10,21] The conflicting results reported by these two groups could be attributed to differences in technique (immunofluorescence microscopy vs cell fractionation) and varying specificities of the APC antibodies used. The simplest explanation for all published results to date is that both NES regions influence the ultimate destination of full-length APC protein, with one region dominating in particular circumstances and the other dominating under other conditions.

APC Shuttles between the Nucleus and Cytoplasm

The nearly concurrent reports from three independent labs that APC could shuttle between the nucleus and cytoplasm left little doubt that there was support for this conclusion. The detailed mechanism and the implications of APC nuclear shuttling have proven the biggest obstacles to reaching complete agreement by active researchers in this field.

If nuclear export of APC is Crm1-dependent, treating cells with LMB should result in an accumulation of APC in the nucleus.[22] Indeed, full-length APC was more abundantly nuclear after LMB treatment in several different cell types, including human cervical carcinoma (HeLa) and human colon carcinoma (HCT116) cells, as determined by cell fractionation.[15,21] Furthermore, Drosophila APC2, while only approximately one-third the size of full-length human APC (Fig. 1B), showed apparent nuclear accumulation following LMB treatment of Drosophila embryos.[19] Truncated forms of human APC also accumulated in the nucleus following LMB-treatment of colorectal cancer cell lines SW480 and HT29, suggesting that the nuclear-cytoplasmic shuttling ability of APC is not lost in some colorectal cancer cells.[21] Localization of exogenously expressed full-length and truncated APC also appeared sensitive to LMB treatment, further supporting the conclusion that APC is capable of nucleo-cytoplasmic shuttling and that nuclear export is a CRM-1 dependent mechanism.[8,10,23]

Factors that Affect the Subcellular Distribution of APC

Sequence similarity between $NLS2_{APC}$ and the well-characterized NLS in SV40 large T antigen led to the recognition that analogous regulation mechanisms may be at work, in particular, the ability to modify NLS activity through flanking phosphorylation sites.[11,24] Indeed it appears that phosphorylation at a potential CK2 site upstream of $NLS2_{APC}$ enhances nuclear import of APC; phosphorylation at a potential PKA site just downstream of $NLS2_{APC}$ inhibits nuclear import (Fig. 2A).[11,24] In addition, simultaneous exposure of cells to PKA agonists and CK2 inhibitors resulted in the anticipated accumulation of endogenous APC in the cytoplasm in both dog kidney epithelial (MDCK) and rat intestinal epithelial (IEC-6) cells. APC can be phosphorylated in vitro by both PKA[25] and CK2,[26] with association of endogenous CK2 and APC demonstrated by co-immunoprecipitation.[26] Furthermore, in mouse keratinocytes where endogenous APC was predominantly nuclear as judged by cell fractionation, treatment with an inhibitor of p38MAPK, a kinase that activates CK2 and inhibits PKA, resulted in a redistribution of APC to the cytoplasm (Fig. 2B).[27] The collective data indicates that nuclear APC levels are regulated, at least in part, by modification of NLS_{APC} activity by phosphorylation.

Other clues regarding the function of APC in the nucleus have been found by examining various conditions that effect localization of APC. The subcellular distribution of both full-length and truncated APC appears relatively constant as cells progress through the cell cycle.[24,28] In contrast, canine kidney (MDCK) and rat intestinal (IEC6) epithelial cells grown to confluence and then maintained at high density for several days still displayed some nuclear APC, but exhibited more cytoplasmic APC than the subconfluent proliferating cells.[24] A similar observation was made using thyroid carcinoma cells (C643), HEK293, C57MG and HCT116 cells, all of which express full-length APC.[28,29] Together, this increase in cytoplasmic APC was visualized using five different

APC antibodies for immunofluorescence microscopy.[24,28,29] The increase in cytoplasmic APC levels only occurred after the cells were maintained in a confluent state for several days, suggesting that the increase was not a direct result of cell-cell contacts and might instead reflect a state of cellular quiescence. The observed increase in cytoplasmic APC was not dependent on sustained nuclear export, as LMB treatment did not induce a nuclear redistribution in the "super-confluent" cells.[24] A fusion protein of a 32 amino acid stretch of APC containing NLS2 (aa. 2028-2058) with β-galactosidase displayed a cell density-influenced distribution similar to endogenous APC, suggesting that NLS2 can regulate this APC redistribution at high cell densities.[24] Furthermore, alterations of the potential phosphorylation sites surrounding NLS2 attenuated this redistribution. Serum starvation of subconfluent MDCK cells also led to an increased number of cells with cytoplasmic APC coincident with decreased cellular proliferation.[28] Taken together, these observations imply that quiescent cells, although they maintain some nuclear APC, have less nuclear import of APC, leaving newly synthesized APC to accumulate in the cytoplasm. Thus, observed increases in cytoplasmic APC as cells stop proliferating and become quiescent likely result from reduced nuclear import, rather than increased export of nuclear APC to the cytoplasm.

Increased levels of cytoplasmic APC in quiescent confluent MDCK cells were reported by a fourth independent lab using a different APC antibody (M-APC). However, the authors argued that because cell fractionation experiments did not confirm this redistribution, the M-APC antibody might recognize a nonspecific nuclear epitope.[21] Another possible interpretation, is that nuclear APC was associated with the nuclear matrix preventing its solubilization so that it was not collected in this particular cell fractionation assay.

Consistent with these observations in wild type cultured cells, variation in APC localization in proliferating versus quiescent cells has also been observed in human colon tissue.[20] In normal epithelia adjacent to the twenty one colon carcinomas analyzed, APC was predominantly cytoplasmic above the crypt (fully differentiated and quiescent cells), but both cytoplasmic and nuclear in the crypt where proliferating cells are located. Truncated APC also displays an increase in cytoplasmic distribution in SW480 colon cancer cells or KAT4 thyroid carcinoma cells grown beyond confluency.[28,29] It was further noted that in large clusters of SW480 cells, peripheral cells displayed more truncated APC in the nucleus, while central cells showed more truncated APC at cell junctions.[28] Both full-length and truncated APC were predominantly nuclear in proliferating cells, implicating nuclear APC in proliferation. Together, the data support a model in which the subcellular localization of APC reflects the proliferative status of the cell, with more cytoplasmic APC found in quiescent cells. Phosphorylation of sites near $NLS2_{APC}$ might participate in control of this redistribution for full-length APC, whereas the truncated forms of APC must rely on other domains to facilitate this change in distribution.

Expanding on the evidence linking APC phosphorylation with cell cycle, full-length APC was found associated with and phosphorylated by casein kinase 2 (CK2) when normal fibroblast cells (TIG-1) were in G_2/M with less association in G_1/S and no association in G_0.[26] This observation is compatible with the model where phosphorylation of APC by CK2 leads to activation of $NLS2_{APC}$ and thus efficient nuclear import of APC. This nuclear import would in turn lead to an overall reduction in the cytoplasmic APC, consistent with a proliferative state (Fig. 2B). The model is further supported by the observation that inhibition of CK2 resulted in relocalization of endogenous APC from the nucleus to the cytoplasm in proliferating cells from two separate epithelial lines.[24]

The binding of APC to CK2 may also provide negative feedback to this system. Two different purified APC fragments (aa. 2086-2394 and 2518-2843) inhibited CK2 activity in vitro.[26] Furthermore, constitutive expression of APC 2086-2394 strongly suppressed proliferation rates in two different cell lines and also impaired colony forming ability in soft agar.[26] If the association of APC with CK2 inhibits the activity of CK2, the resulting reduction in nuclear import of APC might lead to increased cytoplasmic APC, decreased cell proliferation and ultimately cellular quiescence. Because truncated APC typically expressed in colon cancer cells lacks the ability to

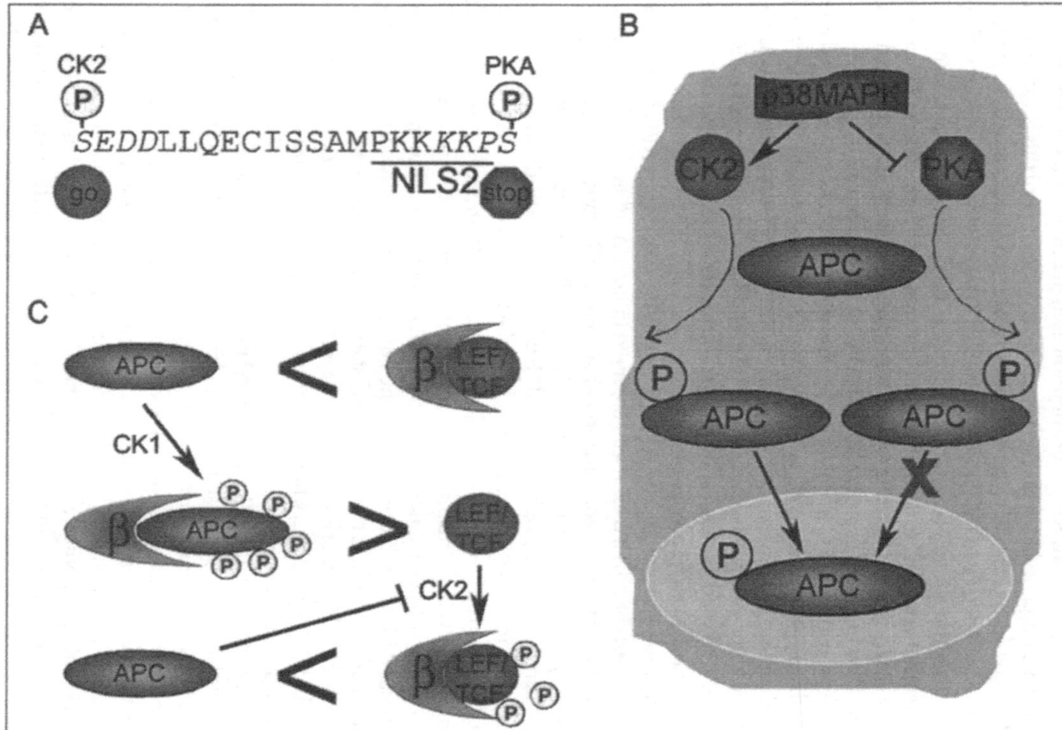

Figure 2. Phosphorylation of APC regulates activity. A) The activity of $NLS2_{APC}$ contributes to nuclear import of APC. While phosphorylation of the serine upstream of NLS2 results in a more active NLS, phosphorylation of the serine just adjacent and downstream of NLS2 inhibits NLS activity. Potential CK2 and PKA recognition sequences are shown in italics. B) p38 MAPK activates CK2 and inhibits PKA. Inhibition of p38MAPK results in reduced levels of nuclear APC. C) LEF/TCF and APC compete for β-catenin binding. β-catenin binds LEF/TCF with a higher affinity than unphosphorylated APC. In contrast, β-catenin binds CK1-phosphorylated APC with a higher affinity than LEF/TCF. CK2-phosphorylated LEF-1 shows stronger binding to β-catenin, but APC can inhibit CK2 activity.

inhibit CK2 activity,[26] these cancer cells would be expected to have impaired abilities to cease proliferation and enter quiescence.

The specific consequences of APC accumulation in the cytoplasm of quiescent cells remain to be defined, but the most obvious impact could be upon β-catenin signaling and cell cycle regulation. Increased levels of cytoplasmic APC could increase destruction complex activity and reduce β-catenin levels by targeting β-catenin for ubiquitin-mediated proteolysis. Nuclear APC might serve a critical role in proliferative cells, perhaps in DNA repair or transcription regulation as discussed below. In quiescent cells, nuclear APC might function in a different capacity, possibly by sequestering nuclear β-catenin from various transcription complexes and thereby inhibiting a Wnt signal, also discussed below.

APC and Nuclear Export of β-Catenin

To date, the best documented function of APC is to oppose a Wnt signal by acting in a complex with Axin, GSK3-β and PP2A to target the protein β-catenin for proteasome-mediated degradation in the cytoplasm in the absence of Wnt signaling.[30] Because this important function of APC is covered in great detail elsewhere in this book (refer to Kennell and Cadigan, this volume) it will be only briefly summarized here. APC contains two types of β-catenin binding domains, the 15 aa and 20 aa repeats (Fig. 1A). The 15 aa repeats (aa. 1020-1169) are sufficient for β-catenin binding, whereas the 20 aa repeats (aa. 1262-2033) bind β-catenin and are required for its destruction. Given the well-characterized role for APC in β-catenin regulation and the knowledge that β-catenin serves as a transcription cofactor in the nucleus, several research groups have examined whether nuclear APC impacts β-catenin localization and activity. This regulation of nuclear β-catenin activity by APC could take the form of one or a combination of the following: enhanced nuclear export of β-catenin, nuclear sequestration of β-catenin, or repression of β-catenin's transcriptional activity.

Initial results implicated nuclear APC in mediating the nuclear export of β-catenin.[8,19,23] In both 3T3 and HCT116 cells, treatment with LMB resulted in a shift of endogenous β-catenin to the nucleus, suggesting that like APC, β-catenin shuttles between nucleus and cytoplasm.[8] Exogenous expression of a truncated APC fragment (aa. 1-1309) which could still bind β-catenin, but couldn't target β-catenin for degradation resulted in a shift of endogenous β-catenin from the nucleus to the cytoplasm of SW480 cells.[8] Blocking nuclear export of this exogenous APC, by either mutation of $NES1_{APC}$ or LMB treatment abolished this cytoplasmic shift of β-catenin, thereby implicating APC in the nuclear export of β-catenin. Similarly, exogenous expression of full-length APC resulted in elimination of both cytoplasmic and nuclear β-catenin in SW480 cells, consistent with the well-characterized role for APC in β-catenin degradation.[23] This reduction of β-catenin levels was abolished by mutation of both $NES1_{APC}$ and $NES2_{APC}$ or by LMB treatment.[23] When nuclear export of APC was inhibited, both APC and the remaining β-catenin were predominantly nuclear. Nuclear export of APC containing nonfunctional $NES1_{APC}$ and $NES2_{APC}$ was restored by addition of a well-characterized NES sequence. This addition also restored APC-mediated degradation of endogenous β-catenin, demonstrating that the loss of APC's export function specifically affected β-catenin export and that the mutation of the N-terminal NESs in APC did not interfere with β-catenin binding or subsequent steps in β-catenin degradation. Together, these results suggest expression of APC with functional N-terminal NESs enhances nuclear export of β-catenin.

The 20R NESs of APC have also been implicated in nuclear export of β-catenin.[19] When expressed in SW480 colon cancer cells, fragments of APC containing 20 aa repeats, SAMP repeats and the 20R NESs (aa. 1342-2075 or aa. 1379-2080) were sufficient to target endogenous β-catenin for destruction[31,19] and localized predominantly to the cytoplasm.[19] LMB treatment or mutation of all three 20R-linked NESs increased the nuclear localization of the APC fragment and resulted in higher β-catenin levels.[19,20] Nuclear β-catenin can serve as a transcription cofactor when bound to transcription factor LEF/TCF family members. Whereas expression of APC (aa 1379-2080) in SW480 cells resulted in reduced β-catenin transcriptional activity as measured using a standard reporter assay, mutation of one, two, or all three 20R-linked NESs resulted in partial restoration of the β-catenin transcription activity.[19] Together, these results are consistent with a role for APC in mediating the nuclear export of β-catenin. However, it is likely that nuclear export of β-catenin can occur by multiple pathways since there is a solid body of evidence for nuclear export of β-catenin independent of APC and Crm1.[14,32-34]

APC and Nuclear Sequestration of β-Catenin

Many of the observations implicating APC in nuclear export of β-catenin are also consistent with nuclear APC dampening β-catenin activity by sequestration.[23] Blocking nuclear export of exogenous full-length APC by either mutations in $NES1_{APC}$ and $NES2_{APC}$ or LMB treatment resulted in nuclear accumulation of both APC and endogenous β-catenin in SW480 colon cancer cells.[23] Other components of the β-catenin destruction complex have been reported to be in the nucleus, with Axin also capable of nuclear export.[35,36] However, since increasing levels of nuclear APC did not result in an apparent reduction in nuclear β-catenin, nuclear APC does not appear to target nuclear β-catenin for destruction within the nuclear compartment.[23] It is of course possible that other components of the destruction complex are limiting in the nucleus so that excess APC cannot reduce β-catenin in this compartment. The unexpected finding that this abundant nuclear β-catenin was not active in standard β-catenin/LEF-1 activity assays led to the proposal that binding of nuclear β-catenin to APC sequesters β-catenin from LEF-1 and renders β-catenin inactive as a transcription cofactor.[23] Because APC and LEF-1 bind to the same domain of β-catenin,[37] it was predicted that the binding of β-catenin to either protein is mutually exclusive.

Indeed, binding affinities of full-length β-catenin to an APC fragment containing 20R3 (aa. 1484-1528) that was homogenously phosphorylated by CK1 and GSK-3β in vitro or to full-length LEF-1 as determined by isothermal titration calorimetry are 10 nM and 20 nM, respectively.[38] These binding affinities indicate that APC might effectively compete with LEF-1 for β-catenin binding, at least when APC is phosphorylated. Nonphosphorylated APC-R3 showed a reduced affinity to β-catenin by three orders of magnitude compared with phosphorylated APC-R3,[38] thus providing a second example where APC phosphorylation influences the function of APC in the nucleus (Fig. 2C). A larger APC fragment containing 20R2 and 20R3 (aa. 1362-1540) also showed a 300-500 fold increase in binding affinity to β-catenin when APC was phosphorylated.[39] Furthermore, this same phosphorylated APC fragment displaced TCF-3 bound to β-catenin in an in vitro binding assay, competed with a LEF-1/DNA complex for β-catenin binding in a mobility shift assay and inhibited β-catenin/LEF-1 mediated transcription in vitro.[40] While collectively these measurements reveal that phosphorylated APC competes effectively with LEF/TCF for β-catenin binding, recent data suggests phosphorylation of LEF-1 by CK2 also enhances LEF-1 binding to β-catenin.[41]

Together, these data support a model for regulation of Wnt signaling in the nucleus by means of a competition between APC and LEF-1 for β-catenin binding that is dependent on the phosphorylation state of APC and LEF-1 (Fig. 2C). β-catenin interaction with phosphorylated APC is favored over a β-catenin/LEF-1 interaction. CK2 phosphorylation of LEF-1 increases the affinity of LEF-1 for β -catenin. APC binds to CK2 and renders it inactive, leaving LEF-1 in an unphosphorylated form and thereby less able to bind β-catenin. If nuclear APC functions by binding and sequestering β-catenin from the transcription complex, APC could very quickly dampen a Wnt signal once the Wnt ligand was no longer bound to its receptor. Nuclear export and subsequent cytoplasmic degradation of β-catenin would not be required for an immediate cellular response to removal of Wnt ligand. Consistent with this model, truncated APC found in most colon cancers binds to CK2 but can not inactivate it, thus leaving CK2 able to phosphorylate LEF-1. Truncated APC would not readily compete with the CK2-phosphorylated LEF-1 for β-catenin binding, leading to constitutive Wnt signaling in the nucleus of the colon cancer cells.

Consistent with a sequestration function of nuclear APC, endogenous APC and β-catenin co-immunoprecipitate from nuclear lysates of HCT116 colon cancer cells, using antibodies to either protein.[23] Full-length APC expressed in SW480 cells in the presence of the proteasome-inhibiting drug MG132 colocalized with endogenous β-catenin in the nucleus, the cytoplasm and along the cytoskeleton, consistent with the notion that APC and β-catenin can interact in the nucleus.[8] Moreover, whereas ectopic expression of LEF-1 in mouse NIH 3T3 fibroblasts led to β-catenin accumulation, co-expression of truncated APC (1-1309) which can bind β-catenin but not target it for destruction, reduced this LEF-1-induced increase in β-catenin levels.[34] Additionally, in SW480 cells, ectopic expression of truncated APC (1-1309) also led to a reduction in β-catenin activity, even when this APC had a nonfunctional NES1. Similarly, expression of a GFP-tagged APC (aa. 1379-2080) with inactivating mutations in all three 20R-linked NESs still reduced endogenous β-catenin activity in SW480 colon cancer cells by half, suggesting that nuclear APC can influence β-catenin activity in the absence of nuclear export.[20]

Initially reported as evidence against the sequestration model, SW480 cells showed a redistribution of otherwise predominantly nuclear β-catenin to the cytoplasm when APC (aa. 1-1309) was expressed ectopically.[34] This cytoplasmic redistribution was blocked by further co-expression of LEF-1, suggesting APC could not compete effectively with LEF-1 for β-catenin binding. In light of the recent demonstration that full-length but not truncated APC can inhibit CK2, thereby reducing LEF-1's ability to bind β-catenin, the data gathered in the SW480 cells is consistent with the sequestration model. Therefore, APC binding to nuclear β-catenin and sequestering it from the LEF-1/TCF transcription complex remains a viable mechanism by which APC counteracts β-catenin. Moreover, recent evidence implicates nuclear retention rather than nuclear export as the dominant mechanism by which nuclear APC regulates nuclear β-catenin.[42] Fluorescence recovery after photobleach (FRAP) of YFP-tagged β-catenin expressed in HEK293 cells, revealed highly efficient shuttling of β-catenin between nucleus and cytoplasm.[42] Co-expression of APC with YFP-tagged β-catenin slowed down this shuttling rather than accelerated it, suggesting that regulation of nuclear β-catenin by APC involves nuclear retention rather than active nuclear export.

The Role of Other Nuclear Proteins in APC Mediated β-Catenin Sequestration

There is growing evidence that other nuclear proteins might bind APC and modulate the ability of APC to displace β-catenin from the LEF-1/TCF complex. Transcription factor activator protein AP-2α coprecipitates with full-length and truncated APC in three different cell lines[43] and two different APC fragments (aa. ~1-400 and ~400-1000) could bind AP-2α in a GST pull down assay. Overexpression of AP-2α in HEK293, HCT116 or HT29 cells appeared to decrease the association between TCF-4 and β-catenin and decreased β-catenin activity. The fact that overexpressed AP-2α did not associate with TCF-4, nor did it change the overall levels of nuclear β-catenin, suggests a role for AP-2α in APC-mediated sequestration of nuclear β-catenin. The exact mechanism behind this modulation has yet to be determined.

The transcriptional corepressor C-terminal binding protein (CtBP) may also affect APC-mediated β-catenin sequestration. CtBP was identified by mass spectrometry as a protein in crude Drosophila embryonic extract associated with Drosophila APC2-GST.[44] Direct binding between human APC (aa. 918-1698) and CtBP was also demonstrated by GST-pull down assays and further, endogenous APC and CtBP were coprecipitated from 293T, HCT116 and SW480 cell lysates.[44] A more recent report contradicts some of the initial findings regarding CtBP and APC interactions and showed that although full-length APC from human epithelial kidney cells (HEK-293) coprecipitated with endogenous CtBP, truncated APC representing roughly the N-terminal half of APC did not coprecipitate with CtBP in either of two colon cancer cell lines (HT29 and SW480).[40] This casts doubt as to the precise APC domain responsible for CtBP binding. To test for the relevance of the APC-CtBP interaction to nuclear ß-catenin function, the sequestration of nuclear β-catenin was forced by expressing APC linked to an additional NLS resulting in a decrease in β-catenin activity. This decrease in activity was partially attenuated by alanine substitutions in the putative CtBP binding domain of APC, consistent with CtBP-binding enhancing the ability of APC to sequester nuclear β-catenin. Given the known functions of CtBP, one might predict that APC delivers CtBP to the β-catenin/TCF complex where CtBP represses

TCF activity. However, CtBP did not appear to coprecipitate with TCF-4 in the mammalian cells examined.[44]

Examination of the regulation of c-myc, a target gene downstream of Wnt signaling, has helped to shed light on the mechanisms involved in APC-CtBP nuclear function.[40] To examine the dynamic nature of proteins bound to the *c-myc* enhancer element, various transcription coregulators were immunoprecipitated from mouse myoblast cells (C2C12) treated with lithium chloride (an inhibitor of GSK3β in the destruction complex) to increase β-catenin levels. The chromatin immunoprecipitation (ChIP) assay demonstrated a transient association of both β-catenin and APC with c-myc enhancer DNA accompanying this increase in β-catenin levels. Surprisingly, this study also showed that truncated APC associated with the *c-myc* enhancer in HT29 cells. Induction of full-length APC in HT29 cells resulted in a transient association of full-length APC with the c-myc enhancer DNA. A similar temporal association with the c-myc enhancer was shown for two other transcriptional corepressors, CtBP and Yin yang 1 (YY1), as well as for β-transducing repeat-containing protein (βTrCP) which has been shown to target β-catenin for ubiquitination and subsequent degradation. The authors suggested that binding of the APC/βTrCP/CtBP/YY1 complex to the c-myc enhancer leads to rapid repression of *c-myc* transcription. Of note in this set of experiments, APC was associated with DNA at times when β-catenin was not, revealing that APC might regulate transcription by multiple mechanisms, not merely by displacing β-catenin from the LEF-1/TCF complex. Overall, these experiments indicated a similar temporal association of APC and transcription corepressors with the chromatin of a Wnt-responsive gene. Thus, it is likely that APC opposes a Wnt signal not only by targeting cytoplasmic β-catenin for destruction, but also by binding and sequestering β-catenin and by associating with corepressors at the transcription complex (Fig. 3). Future experiments will surely clarify the broad range of APC activities as they relate to transcription control.

Interaction of APC with DNA and Other Nuclear Proteins

There is preliminary evidence for a direct interaction between APC and DNA. A 180 kDa protein recognized by antibodies raised against both the N- and C-terminal region of APC was identified in the soluble nuclear fraction of both HEK-293 and HCT116 cells.[45] Although it is puzzling that an APC truncation would retain both N- and C-terminal regions, three recombinant protein fragments, all from the C-terminal half of APC, bound plasmid DNA in mobility shift assays. Repetitive preparative mobility shift assays using two of these APC fragments revealed no consensus DNA binding sequence, but did suggest a higher affinity for A/T nucleotides vs G/C.

In addition to DNA, other nuclear proteins can interact with APC. Full length APC from HCT116 colon cancer cells can bind polymerase β and PCNA, part of the base excision repair (BER) pathway, raising the intriguing possibility that APC participates in DNA repair.[46] The APC fragment (aa. 1250-1269) capable of polymerase β and PCNA binding contains a PCNA-interacting protein-like box (PIP-like box). The PIP-like domain of APC appeared to block repair of a multi-nucleotide repair patch, "long patch BER", but not a damaged single nucleotide base, "short patch BER", in vitro.

Furthermore, APC also interacts with protein tyrosine phosphatase (PTP-BL) in the nucleus. The C-terminal 19 amino acids of APC appear to mediate binding to PTP-BL, with a dissociation constant of 8 nM as measured using surface plasmon resonance analysis.[47] APC and PTP-BL colocalized both at the tips of cellular extensions and in the nucleus of MDCK cells, once again implicating phosphorylation as a means to regulate nuclear function(s) of APC. The exact roles of these additional nuclear interactions of APC and whether they impinge upon Wnt signal transduction will require further investigation.

The Requirement for Nuclear APC in Tumor Suppression

There is limited evidence supporting a role for nuclear APC in tumor suppression in the colon. The vast majority of germline and sporadic APC mutations in colorectal cancers result in protein

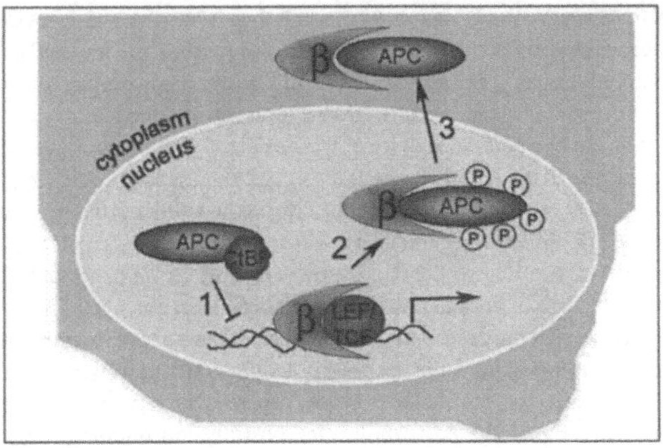

Figure 3. Model for the roles of nuclear APC in regulation of Wnt target genes. APC modifies expression of Wnt target genes by three mechanisms. 1) APC associates with transcription corepressors, such as CtBP, potentially guiding them to the transcription complex. 2) Phosphorylated APC competes with LEF/TCF for β-catenin binding, thus displacing β-catenin from the transcription complex. 3) APC may assist in the nuclear export of β-catenin for cytoplasmic degradation.

truncation, leaving only the N-terminal half of the APC protein expressed. Although lacking the central NLSs and NESs, this truncated APC protein typically retains the ability to shuttle and thus could negatively impact nuclear processes that depend on the C-terminus of APC. In human colon tissue, truncated APC protein as detected with N-terminal APC antibodies remained strongly nuclear in polyps and carcinomas.[6] However, the frequency of cytoplasmic APC progressively increased, with 60% of the carcinoma tissues showing cytoplasmic APC compared to only 4% of the normal tissues sampled.[6] Since inactivating *APC* mutations are found in over 80% of all colon carcinomas, most carcinoma tissue should display no full-length APC. Indeed, of the 148 images of human carcinoma tissue analyzed, only 11% showed nuclear staining using C-terminal APC antibodies.[6] Although they represent a minority, there are several germline missense mutations found in *APC*. Of note, a few of these mutations result in amino acid substitutions in or near the NLS and NES domains. While the missense mutations indicate a potential requirement for these domains in APC-mediated tumor suppression, to date, none of these amino acid substitutions have been examined for their impact on NLS and NES activity. Overall, the changes in APC localization as human colon tissue progresses from normal to polyp to carcinoma and the germline mutations found in and near APC's nuclear import and export signals are consistent with a role for nuclear APC in tumor suppression.

In addition to colon tissue, nuclear APC has been described in alveolar soft-part sarcoma and in over 62% of 92 small cell lung cancers (SCLC) sampled.[48,49] Besides the human and mouse studies, nuclear APC has been described in cells from worm,[50] frog,[51] fruit fly,[19] monkey[52] and dog.[24] At present, mouse models have not been particularly informative in determining the nuclear function of APC. Mice expressing only a truncated form of APC (aa. 1-1638) had reduced viability, but those surviving to adulthood had no increase in intestinal polyp number compared to wild-type mice.[53] This surprising finding is consistent with the observation that APC-interacting proteins β-tubulin, EB-1 and discs large (DLG) retained normal subcellular localization patterns in these mice and in fact, the truncated APC remained nuclear. The ARM domain still retained in this truncated APC likely contributed to nuclear import. The definitive genetic test for nuclear function of APC in a model organism has yet to be completed, but entails abrogation of the nuclear import ability of APC in a mouse model.

Future Applications, New Research, and Anticipated Developments

Although there are many areas of consensus in the field of nuclear APC localization and function, there have also been some conflicting results with no apparent resolution. Experiments performed with protein fragments of APC over-expressed in a few colon cancer cell lines have led to speculation about the relative activity of various NESs and NLSs. One complication in interpreting such results is that functions ascribed to small fragments of APC might not accurately reflect functions when these sequences are in the context of full-length native APC. In addition, these exogenous APC fragments are often over-expressed compared to endogenous APC levels and thus might saturate binding to protein partners in limited supply. Finally, there appears to be no readily available normal human colon epithelial cell line, nor a cell line completely lacking APC. In the absence of such model cell lines, most experiments are performed using colon cancer cell lines that express endogenous APC, either full-length or truncated. Because truncated APC still retains an N-terminal oligomerization domain,[54] there is the potential for heterodimerization between full-length and truncated APC molecules, possibly confounding results.

Unfortunately, complexity is also encountered when endogenous APC is examined. The specificity of several APC antibodies has been questioned, with no clear consensus about the "best" antibody to use for APC detection.[21,55-57] With the potential specificity issues surrounding APC detection by immunofluorescence microscopy, one might think that cellular fractionation followed by immunoblot would serve as a better "gold standard" for APC distribution. Regrettably, use of different fractionation protocols leads to considerable variation in the distribution of soluble nuclear proteins and cytoskeletal/membrane associated proteins and potential elimination of nuclear matrix-associated insoluble proteins, making cell fractionation followed by immunoblot analysis potentially problematic as well. The best solution appears to be analysis of APC using several complimentary techniques and a variety of APC antibodies with the least amount of demonstrated nonspecific cross reactivity, with quantitation whenever possible.

This is an exciting time for researchers interested in nuclear functions of APC. APC has been identified in the nucleus of many different cultured cell lines as well as in normal human colon tissue and tissue from other organisms. Intrinsic localization signals have been characterized that facilitate shuttling of APC between the nucleus and cytoplasm. Phosphorylation of APC provides one underlying mechanism by which APC localization and activity are regulated. APC distribution within a cell appears to correlate with the proliferative or quiescent status of the cell. Several intriguing nuclear binding partners have already been identified, with consequences of APC binding to these nuclear proteins in preliminary stages. For example, hints of a direct link between APC and the transcription complex or the DNA repair machinery are beginning to emerge. APC likely serves to regulate Wnt signaling, in part, by competing with nuclear LEF-1 for β-catenin binding. It is anticipated that future work will clarify the role of APC in association with various proteins at the c-myc enhancer and potentially as APC interacts directly with DNA. Future biochemical analysis is expected to reveal additional nuclear proteins that interact with APC, as well as the cellular consequences of APC association with these proteins. The definitive test to assess the requirement of nuclear APC in suppression of colorectal tumors will be to eliminate nuclear APC from a model organism. One such mouse model currently in production should prove useful to address some of the remaining questions. Ultimately, a more thorough understanding of the many functions of the tumor suppressor APC will ultimately lead to better diagnostics and more targeted therapeutics for colorectal cancer.

Acknowledgements

The author would like to thank Drs. T. Chris Gamblin, Audrey Lamb and Erik Lundquist for critical review of the manuscript and Jamie Cunningham for contributions to Figure 1. The author also thanks Drs. Brooke McCartney and Inke Nathke for soliciting this chapter and for their helpful editorial suggestions. The author wishes to acknowledge the support of the National Cancer Institute/National Institute of Health (RO1CA109220-01). Apologies are extended to colleagues whose research was omitted from the review due to size limitations.

References

1. Neufeld KL, White RL. Nuclear and cytoplasmic localizations of the adenomatous polyposis coli protein. Proc Natl Acad Sci USA 1997; 94:3034-3039.
2. Munemitsu S, Souza B, Muller O, Albert I, Rubinfeld B, Polakis P. The APC gene product associates with microtubules in vivo and promotes their assembly in vitro. Cancer Research 1994; 54(14):3676-3681.
3. Smith K, Levy D, Maupin P, Pollard T, Vogelstein B, Kinzler K. Wild-type but not mutant APC associates with the microtubule cytoskeleton. Cancer Research 1994; 54(14):3672-3675.
4. Nathke IS, Adams CL, Polakis P, Sellin JH, Nelson J. The adenomatous polyposis coli tumor suppressor protein localizes to plasma membrane sites involved in active cell migration. J Cell Biol 1996; 134:165 -179.
5. Wong MH, Hermiston ML, Syder AJ, Gordon JI. Forced expression of the tumor suppressor adenomatosis polyposis coli protein induces disordered cell migration in the intestinal epithelium. Proc Natl Acad Sci USA 1996; 93(18):9588-9593.
6. Anderson CB, Neufeld KL, White RL. Subcellular distribution of Wnt pathway proteins in normal and neoplastic colon. Proc Natl Acad Sci USA Jun 25 2002; 99(13):8683-8688.
7. Sena P, Saviano M, Monni S, Losi L, Roncucci L, Marzona L, Pol AD. Subcellular localization of beta-catenin and APC proteins in colorectal preneoplastic and neoplastic lesions. Cancer Lett Sep 28 2006; 241(2):203-212.
8. Henderson B. Nuclear-cytoplasmic shuttling of APC regulates β-catenin subcellular localization and turnover. Nat Cell Biol 2000; 2:653-660.
9. Paine PL, Moore LC, Horowitz SB. Nuclear envelope permeability. Nature Mar 13 1975; 254(5496):109-114.
10. Galea MA, Eleftheriou A, Henderson BR. ARM domain-dependent nuclear import of adenomatous polyposis coli protein is stimulated by the B56 alpha subunit of protein phosphatase 2A. J Biol Chem 2001; 276(49):45833-45839.
11. Zhang F, White R, Neufeld K. Phosphorylation near nuclear localization signal regulates nuclear import of APC protein. Pro. Natl Acad USA 2000; 97:12577-12582.
12. Koike M, Kose S, Furuta M, Taniguchi N, Yokoya F, Yoneda Y, Imamoto N. beta-Catenin shows an overlapping sequence requirement but distinct molecular interactions for its bidirectional passage through nuclear pores. J Biol Chem Aug 6 2004; 279(32):34038-34047.
13. Yokoya F, Imamoto N, Tachibana T, Yoneda Y. beta-catenin can be transported into the nucleus in a Ran-unassisted manner. Mol Biol Cell 1999; 10(4):1119-1131.
14. Wiechens N, Fagotto F. CRM1- and Ran-independent nuclear export of beta-catenin. Curr Biol 2001; 11(1):18-27.
15. Neufeld KL, Nix DA, Bogerd H, Kang Y, Beckerle MC, Cullen BR, White RL. Adenomatous Polyposis Coli Protein Contains Two Nuclear Export Signals and Shuttles between the Nucleus and Cytoplasm. Proc Natl Acad Sci USA 2000; 97:12085-12090.
16. Tickenbrock L, Cramer J, Vetter IR, Muller O. The coiled coil region (amino acids 129-250) of the tumor suppressor protein adenomatous polyposis coli (APC). Its structure and its interaction with chromosome maintenance region 1 (Crm-1). J Biol Chem Aug 30 2002; 277(35):32332-32338.
17. McCartney BM, Dierick HA, Kirkpatrick C, Moline MM, Baas A, Peifer M, Bejsovec A. Drosophila APC2 is a cytoskeletally-associated protein that regulates wingless signaling in the embryonic epidermis. J Cell Biol 1999; 146(6):1303-1318.
18. Yu X, Waltzer L, Bienz M. A new Drosophila APC homologue associated with adhesive zones of epithelial cells. Nat Cell Biol 1999; 1(3):144-151.
19. Rosin-Arbfeld R, Townsley F, Bienz M. The APC tumour suppressor has a nuclear export function. Nature 2000; 406:1009-1012.
20. Rosin-Arbesfeld R, Cliffe A, Brabletz T, Bienz M. Nuclear export of the APC tumour suppressor controls beta-catenin function in transcription. EMBO J Mar 3 2003; 22(5):1101-1113.
21. Brocardo M, Nathke IS, Henderson BR. Redefining the subcellular location and transport of APC: new insights using a panel of antibodies. EMBO Rep Feb 2005; 6(2):184-190.
22. Fukuda M, Asano S, Nakamura T, Adachi M, Yoshida M, Yanagida M, Nishida E. CRM1 is responsible for intracellular transport mediated by the nuclear export signal. Nature 1997; 390(6657):308-311.
23. Neufeld KL, Zhang F, Cullen BR, White RL. APC-mediated down-regulation of β-Catenin activity involves nuclear sequestration and nuclear export. EMBO Rep 2000; 6:519-523.
24. Zhang F, White RL, Neufeld KL. Cell density and phosphorylation control the subcellular localization of adenomatous polyposis coli protein. Mol Cell Biol 2001; 21(23):8143-8156.
25. Rubinfeld B, Albert I, Porfiri E, Fiol C, Munemitsu S, Polakis P. Binding of GSK3β to the APC-β-catenin complex and regulation of complex assembly. Science 1996; 272:1023-1026.

26. Homma MK, Li D, Krebs EG, Yuasa Y, Homma Y. Association and regulation of casein kinase 2 activity by adenomatous polyposis coli protein. Proc Natl Acad Sci USA Apr 30 2002; 99(9):5959-5964.
27. Hildesheim J, Salvador JM, Hollander MC, Fornace AJ, Jr. Casein kinase 2- and protein kinase A-regulated adenomatous polyposis coli and beta-catenin cellular localization is dependent on p38 MAPK. J Biol Chem Apr 29 2005; 280(17):17221-17226.
28. Fagman H, Larsson F, Arvidsson Y, Meuller J, Nordling M, Martinsson T, Helmbrecht K, Brabant G, Nilsson M. Nuclear accumulation of full-length and truncated adenomatous polyposis coli protein in tumor cells depends on proliferation. Oncogene Sep 4 2003; 22(38):6013-6022.
29. Davies ML, Roberts GT, Spiller DG, Wakeman JA. Density-dependent location and interactions of truncated APC and beta-catenin. Oncogene Feb 19 2004; 23(7):1412-1419.
30. Giles RH, van Es JH, Clevers H. Caught up in a Wnt storm: Wnt signaling in cancer. Biochim Biophys Acta Jun 5 2003; 1653(1):1-24.
31. Munemitsu S, Albert I, Souza B, Rubinfeld B, Polakis P. Regulation of intracellular beta-catenin levels by the adenomatous polyposis coli (APC) tumor-suppressor protein. Proc Natl Acad Sci USA 1995; 92(7):3046-3050.
32. Eleftheriou A, Yoshida M, Henderson BR. Nuclear export of human beta-catenin can occur independent of CRM1 and the adenomatous polyposis coli tumor suppressor. J Biol Chem Jul 13 2001; 276(28):25883-25888.
33. Hendriksen J, Fagotto F, van der Velde H, van Schie M, Noordermeer J, Fornerod M. RanBP3 enhances nuclear export of active (beta)-catenin independently of CRM1 J Cell Biol. Dec 5 2005; 171(5):785-797.
34. Henderson BR, Galea M, Schuechner S, Leung L. Lymphoid enhancer factor-1 blocks adenomatous polyposis coli-mediated nuclear export and degradation of beta-catenin. Regulation by histone deacetylase 1. J Biol Chem Jul 5 2002; 277(27):24258-24264.
35. Cong F, Varmus H. Nuclear-cytoplasmic shuttling of Axin regulates subcellular localization of beta-catenin. Proc Natl Acad Sci USA Mar 2 2004; 101(9):2882-2887.
36. Wiechens N, Heinle K, Englmeier L, Schohl A, Fagotto F. Nucleo-cytoplasmic shuttling of Axin, a negative regulator of the Wnt-beta-catenin Pathway. J Biol Chem Feb 13 2004; 279(7):5263-5267.
37. Orsulic S, Huber O, Aberle H, Arnold S, Kemler R. E-cadherin binding prevents beta-catenin nuclear localization and beta-catenin/LEF-1-mediated transactivation. J Cell Sci 1999; 112(Pt 8):1237-1245.
38. Choi HJ, Huber AH, Weis WI. Thermodynamics of beta-catenin-ligand interactions: the roles of the N- and C-terminal tails in modulating binding affinity. J Biol Chem Jan 13 2006; 281(2):1027-1038.
39. Xing Y, Clements WK, Le Trong I, Hinds TR, Stenkamp R, Kimelman D, Xu W. Crystal structure of a beta-catenin/APC complex reveals a critical role for APC phosphorylation in APC function. Mol Cell Aug 27 2004; 15(4):523-533.
40. Sierra J, Yoshida T, Joazeiro CA, Jones KA. The APC tumor suppressor counteracts beta-catenin activation and H3K4 methylation at Wnt target genes. Genes Dev Mar 1 2006; 20(5):586-600.
41. Wang S, Jones KA. CK2 controls the recruitment of Wnt regulators to target genes in vivo. Curr Biol Nov 21 2006; 16(22):2239-2244.
42. Krieghoff E, Behrens J, Mayr B. Nucleo-cytoplasmic distribution of beta-catenin is regulated by retention. J Cell Sci Apr 1 2006; 119(Pt 7):1453-1463.
43. Li Q, Dashwood RH. Activator protein 2alpha associates with adenomatous polyposis coli/beta-catenin and Inhibits beta-catenin/T-cell factor transcriptional activity in colorectal cancer cells. J Biol Chem Oct 29 2004; 279(44):45669-45675.
44. Hamada F, Bienz M. The APC tumor suppressor binds to C-terminal binding protein to divert nuclear beta-catenin from TCF. Dev Cell Nov 2004; 7(5):677-685.
45. Deka J, Herter P, Sprenger-Haussels M, Koosch S, Franz D, Muller KM, Kuhnen C, Hoffmann I, Muller O. The APC protein binds to A/T rich DNA sequences. Oncogene 1999; 18(41):5654-5661.
46. Narayan S, Jaiswal AS, Balusu R. Tumor suppressor APC blocks DNA polymerase beta-dependent strand displacement synthesis during long patch but not short patch base excision repair and increases sensitivity to methylmethane sulfonate. J Biol Chem Feb 25 2005; 280(8):6942-6949.
47. Erdmann KS, Kuhlmann J, Lessmann V, Herrmann L, Eulenburg V, Muller O, Heumann R. The Adenomatous Polyposis Coli-protein (APC) interacts with the protein tyrosine phosphatase PTP-BL via an alternatively spliced PDZ domain. Oncogene 2000; 19(34):3894-3901.
48. Kuhnen C, Herter P, Monse H, Kahmann S, Muehlberger T, Vogt PM, Steinau HU, Muller KM, Muller O. APC and beta-catenin in alveolar soft part sarcoma (ASPS)--immunohistochemical and molecular genetic analysis. Pathol Res Pract 2000; 196(5):299-304.
49. Kotsinas A, Evangelou K, Zacharatos P, Kittas C, Gorgoulis VG. Proliferation, but not apoptosis, is associated with distinct beta-catenin expression patterns in non-small-cell lung carcinomas: relationship with adenomatous polyposis coli and G(1)-to S-phase cell-cycle regulators. Am J Pathol Nov 2002; 161(5):1619-1634.

50. Mizumoto K, Sawa H. Cortical beta-catenin and APC regulate asymmetric nuclear beta-catenin localization during asymmetric cell division in C. elegans. Dev Cell Feb 2007; 12(2):287-299.
51. Klymkowsky MW, Williams BO, Barish GD, Varmus HE, Vourgourakis YE. Membrane-anchored plakoglobins have multiple mechanisms of action in Wnt signaling. Mol Biol Cell 1999; 10(10):3151-3169.
52. Langford KJ, Lee T, Askham JM, Morrison EE. Adenomatous polyposis coli localization is both cell type and cell context dependent. Cell Motil Cytoskeleton Aug 2006; 63(8):483-492.
53. Smits R, Kielman MF, Breukel C, Zurcher C, Neufeld K, Jagmohan-Changur S, Hofland N, van Dijk J, White R, Edelmann W, Kucherlapati R, Khan PM, Fodde R. Apc1638T: a mouse model delineating critical domains of the adenomatous polyposis coli protein involved in tumorigenesis and development. Genes Dev 1999; 13(10):1309-1321.
54. Joslyn G, Richardson D, White R, Alber T. Dimer formation by an N-terminal coiled coil in the APC protein. Proc Nat Acad Sci USA 1993; 90:11109-11113.
55. Roberts GT, Davies ML, Wakeman JA. Interaction between Ku80 protein and a widely used antibody to adenomatous polyposis coli. Br J Cancer Jan 27 2003; 88(2):202-205.
56. Rosin-Arbesfeld R, Ihrke G, Bienz M. Actin-dependent membrane association of the APC tumour suppressor in polarized mammalian epithelial cells. EMBO J Nov 1 2001; 20(21):5929-5939.
57. Davies ML, Roberts GT, Stuart N, Wakeman JA. Analysis of a panel of antibodies to APC reveals consistent activity towards an unidentified protein. Br J Cancer 2007; 97(3):384-390.

Chapter 3

APC in Cell Migration

Sandrine Etienne-Manneville*

Abstract

Adenomatous Polyposis Coli (APC) is a tumor suppressor protein involved in the initiation and progression of colon cancer. The most widely accepted function of APC is to participate to the Wnt signaling pathway, by downregulating β-catenin and thereby controlling gene transcription and cell proliferation. However, APC is clearly a multifunctional protein whose loss contributes to tumor formation in multiple ways. Regulation of APC localization during cell migration and the ability of APC to bind multiple polarity proteins and microtubule-associated molecules support the idea that APC plays a key role in directed cell migration and that this function may contribute to its tumour suppressor activity.

Introduction

The *adenomatous polyposis coli* (*Apc*) gene is a tumour suppressor involved in familial adenomatous polyposis and in sporadic colorectal tumours. Mutation in *Apc* occurs early during the progression of colorectal cancer suggesting that disruption of APC's functions is a key event in the initiation of colon adenomas.[1] Since these tumours arise from the epithelial layer, APC function is probably crucial for the maintenance of epithelial structure. Cell renewal within the intestinal epithelium is tightly regulated and position-dependent. Absorptive cells generated near the bottom of the intestinal crypt migrate in tight cohorts towards the upper regions of the villi until they either die or are shed into the gut lumen.[2] APC expression in the gut epithelium typically increases from the base of the crypt to the top of the villi extending into the lumen and therefore is associated with differentiation and directed migration of epithelial cells.[3] Whether *Apc* mutations arise in stem cells near the bottom of the crypt or in migrating cells is still debated[4] (see Kwong and Dove, this volume). Interestingly, in adenomas, cells do not to migrate toward the lumen but display abnormal migration properties,[4,5] suggesting a link between APC and cell migration.

The *Apc* gene encodes a large protein with multiple cellular functions and interactions, including its well-characterized function in signal transduction in the Wnt-signalling pathway (Fig. 1). Loss of *Apc* or truncation of the Axin/β-catenin binding domain results in the inappropriate stabilization of β-catenin and consequently an increase in the transcription of Wnt target genes, including those associated with proliferation such as c-myc and cyclin D1.[6] It is widely believed that alterations in the regulation of epithelial cell proliferation and perturbation of the normal repression of proliferation during migration along the crypt play a key role during colonic cell transformation.[7,8] However, several domains in the APC molecule mediate interactions of APC with a number of proteins not involved in Wnt signalling, raising the possibility that APC may play concomitant functions in multiple cellular events, such as directed cell migration (Fig. 1).

The role of APC in cell migration has been confirmed in several in vivo models. Total loss of APC in vivo abrogates cell migration along the crypt-villus axis.[9] Similarly, in *Caenorhabditis*

*Sandrine Etienne-Manneville—Cell Polarity and Migration Group, Institut Pasteur CNRS URA 2582, 25 Rue du Dr Roux, 75724 Paris cedex15, France.
Email: sandrine.etienne-manneville@pasteur.fr

APC Proteins, edited by Inke S. Näthke and Brooke M. McCartney. ©2009 Landes Bioscience and Springer Science+Business Media.

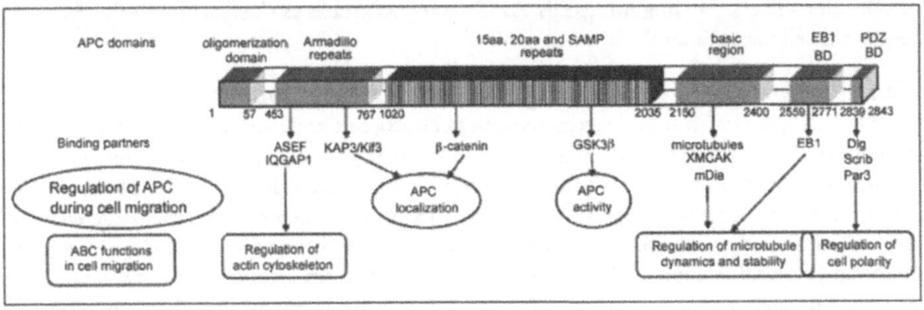

Figure 1. Role of APC domains in cell migration. The human APC protein contains multiple domains involved in protein-protein interactions. Amino acid numbers delimiting each major domain are indicated under the bar. APC binding partners involved in APC regulation during cell migration (round shapes) and in APC functions in directed cell migration (square shapes) are shown.

elegans, inactivation of the APC-related protein APR-1 induces defects in epithelial cell migration.[10] However, in humans, most germ-line mutations in APC do not lead to the total depletion of the protein but to the expression of truncated forms of the molecule. Colorectal carcinogenesis varies in severity depending on the length of the protein resulting from the mutant allele.[11] A majority of germline and somatic mutations in *Apc* occurs in the first half of the coding region and results in the expression of the N-terminal domain of the protein, which is stable and retains its ability to oligomerize. Mice heterozygous for truncated *Apc* after residue 850 (see Fig. 1) develop polyps similar to those of human patients with familial adenomatous polyposis. These polyps lack directed migration of epithelial cells along the crypt-villus axis,[12,13] confirming the essential role of APC in directed migration of epithelial cells and raising the possibility of a dominant negative effect of truncated APC.[14]

Although development of microadenoma correlates with the loss of the wild-type *Apc* allele[15] and high level expression of different truncated APC proteins is not sufficient to induce intestinal polyps or tumours,[16] the role of the mutant APC protein in the early steps of polyp formation is supported by the systematic increase of β-catenin and the decrease of enterocyte migration in histologically normal intestinal mucosa of mice bearing a single germ-line *Apc* mutation.[13] Interestingly, early adenomas in the *Apc*(*Δ716*) polyps are very similar to the normal proliferating cells of the crypt except for the lack of directed migration along the crypt-villus axis,[17] demonstrating that loss of directed migration may be a very early event during tumour development.

In parallel to these in vivo studies, evidence has recently accumulated in cultured cells showing the regulation and the role of APC during cell migration. The first evidence for a potential role of APC in cell migration was the observation that, in motile cells, APC concentrates in the protrusive region.[18] This observation raised three main questions: (1) How is APC recruited to protrusion sites and what is its role there? (2) Does APC affect migration speed or directionality? (3) How does this contribute to tumour formation? This chapter will focus on recent advances made to tackle these questions.

APC Localization in Migrating Cells

Determining the subcellular localization of APC has been quite a challenge and its recruitment to protrusive sites has been debated for a long time. This was due to two major technical limitations: the difficulty in expressing a full-length tagged protein to levels that do not perturb migration and the conflicting results obtained by immuno-localization in fixed cells using different antibodies. Although differences may exist depending on cell type and cellular context,[19] evidence has eventually converged on the idea that in addition to localizing in the nucleus,[20,21] APC localizes at cell-cell contacts.[22] In migrating cells, APC also concentrates as clusters in actively protruding regions of cell membranes.[23]

Actin-Dependent Localization of APC at Cell-Cell Contacts

APC associates with adhesive regions both in Drosophila and mammalian epithelial cells.[24-27] APC has been shown to interact with β-catenin and plakoglobin, two proteins directly interacting with E-cadherins and APC colocalizes with β-catenin at cell-cell contacts.[26,27] However, APC and cadherin binding to β-catenin are mutually exclusive, which suggests that different pools of β-catenin reside near adherens junctions and that APC is not directly associated with the cadherin-β-catenin complex.[28] In fact, perturbation of the actin cytoskeleton and drug wash-out experiments suggest that APC is delivered continuously to the plasma membrane by a dynamic actin-dependent process.[29,30] It is thus possible that APC localizes at cell-cell contacts in part by interacting with an actin-bound protein complex that may contain β- and/or α-catenin. For example, in the early Drosophila embryo, cortical localization of APC2 appears to be dependent on the ß-catenin homologue Armadillo.[31] Alternatively, APC localization may also involve its binding to the actin-binding protein IQGAP-1 (IQ motif-containing GTPase-activating protein 1).[32]

Microtubule-Dependent Localization of APC at Protrusive Sites

APC also localizes at the tip of cell protrusions in various cell types including epithelial cells as well as fibroblasts, astrocytes and neurons.[33-35] In contrast to the cell-cell contact localization, APC localization to protrusive regions of the plasma membrane requires an intact microtubule network and is not affected by actin depolymerization.[33,36] A more precise analysis of APC localization shows that APC forms clusters at microtubule plus-ends[18,23,36,37] (Fig. 2), although APC is also visible along the length of microtubules to a lesser extent. APC mutant proteins that lack the carboxy-terminal third of the molecule do not bind as efficiently as the wild type protein to microtubules,[38] reflecting the presence of at least one microtubule interacting domain in the carboxy-terminal part of APC. In fact, APC includes three regions that can directly or indirectly bind microtubules (Fig. 1).

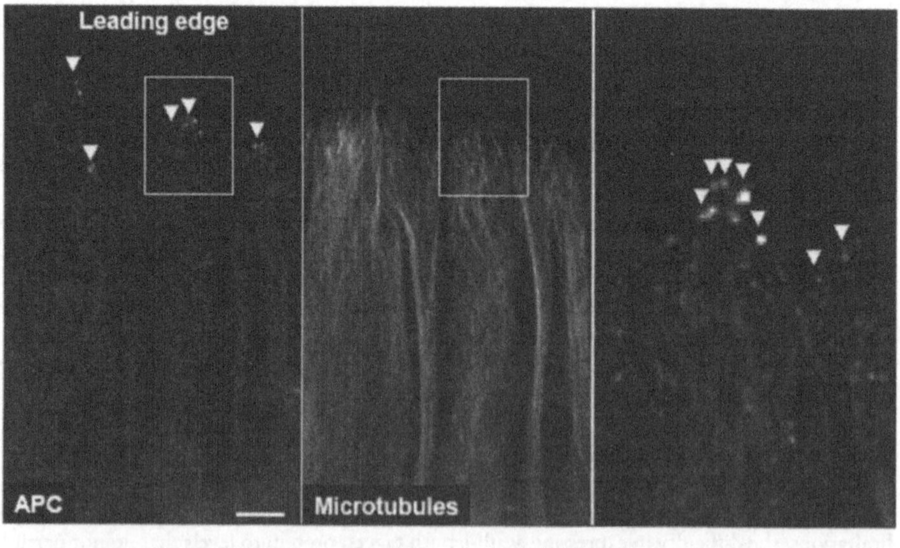

Figure 2. APC localizes at microtubule plus-ends at the leading edge of migrating cells. Migrating astrocytes have been fixed, permeabilized and stained with anti-APC (left panel) and anti-tubulin (middle panel) antibodies. APC forms clusters (arrow heads) which localize at microtubule plus-ends. Bar, 5 μm. Merge image of the boxed area is shown in the right panel.

Two regions involved in APC's interaction with microtubules are located in the carboxy-terminal domain of APC: a basic domain that allows direct microtubule binding[38,39] and a binding site for the microtubule plus-end tracking protein EB1.[40] Expression of the basic domain alone revealed a nonspecific microtubule localization and EB1 binding has been suggested to be involved in APC targeting to microtubule tips.[41] Interestingly, APC clusters do not strictly colocalize with EB1 at microtubule plus-ends.[42] Moreover, expression of the APC binding site of EB1 alone does not disrupt APC clusters in epithelial cells,[42] but does abolish APC recruitment to microtubules in migrating astrocytes.[36] This suggests that EB1 function may vary with cell type or cellular context and points to an additional domain essential for APC localization at microtubule plus-ends.

The third microtubule binding domain of APC is located in the Armadillo repeats and is responsible for APC interaction with microtubule plus-end-directed motor proteins of the kinesin superfamily (KIF), through an association with the kinesin superfamily-associated protein 3 (KAP3).[43] The interaction of APC with KAP3 and Kif3 is required for its accumulation in clusters,[42] leading to the idea that APC is transported towards the plus-ends of microtubules where it oligomerizes to form clusters. Whether APC clusters remain associated with microtubule plus-ends or whether they later associate with actin filaments at the cell cortex where they can remain in absence of microtubules is not clear.

Regulation of APC Localization During Cell Migration

Cell protrusion and migration affect APC localization by promoting its interaction with microtubule plus-ends[33,34,44] and its recruitment to the leading edge. This recruitment requires the kinesin motor Kif3, suggesting that APC shuttles between its different locations via the microtubule network.[45] The interaction between APC and EB1 is regulated by different factors. Protein kinase A (PKA) and p24cdc2 phosphorylate APC and inhibit its interaction with EB1.[41] Conversely, the Rho effector mDia can directly associate with both APC and EB1 and might act as a linker protein to enhance APC-EB1 association.[46]

APC accumulation at the leading edge of migrating cells also requires a spatially regulated GSK3β activity. In Drosophila, APC2 interaction with the cell cortex requires the Drosophila orthologue of GSK3β.[31] GSK3β is a kinase that binds and phosphorylates the β-catenin-APC complex, leading to β-catenin degradation by the proteasome.[47] GSK3β activity is regulated by the Wnt signaling pathway and also by phosphorylation on serine 9, which inhibits its kinase activity. Phosphorylation of GSK3β is also involved in APC recruitment in migrating astrocytes and elongating axons.[33,44] During wound-induced astrocyte migration, GSK3β is phosphorylated via PKCζ downstream of Cdc42 and Par6 leading to APC's association with microtubule plus ends.[33] In developing neurons, NGF promotes PI3K signalling to induce GSK3β phosphorylation by AKT, resulting in APC regulation in the growth cone.[44] However, in differentiating hippocampal neurons, GSK3β inactivation seems to be induced independently of its phosphorylation.[48] How GSK3β regulation leads to APC association with microtubules remains unclear. Since β-catenin is required for the formation of APC membrane clusters,[49] inhibition of GSK3β may induce β-catenin stabilization and indirectly control APC trafficking on microtubules. Alternatively, GSK3β can directly phosphorylate APC and thereby modify its affinity for microtubules[50] and for β-catenin.[51]

In conclusion, APC is concentrated at cell-cell contacts and at microtubule plus-ends. The amount of APC present in these two distinct pools is regulated upon cell migration. The primary cause of cell scattering in invasive carcinomas is the loss of integrity of intracellular junctions, often involving E-cadherin.[52] Although loss of APC does not induce apparent defect in cadherin-based adhesion,[53] APC competition with cadherin for β-catenin binding[54] may somehow control epithelial cell adhesion.[55,56] APC has also been proposed to positively regulate E-cadherin translocation to the cell membrane.[57] Loss of APC at cell-cell junctions may therefore play a role in the initiation of cell migration. More importantly, APC recruitment to the leading edge of migrating cells is likely to either play a role in the generation of the protrusive forces responsible for membrane extension or control intracellular events required for directed cell movement.

APC and Protrusive Activity

The localisation of APC on microtubule plus-ends suggests that APC may directly regulate microtubule dynamics. Moreover, since APC interacts with a wide range of proteins, it is an excellent candidate molecule for controlling essential effectors of cell migration.

APC and Regulation of the Microtubule Cytoskeleton

APC can bind microtubules and affect their dynamics in a number of different ways.[39] APC binds to tubulin subunits with high affinity and promotes microtubule assembly both in vitro and in vivo.[58,59] In addition, APC can bind EB1 to promote microtubule polymerization,[60] possibly by relieving an auto-inhibited conformation of EB1.[61] APC also participates in microtubule stabilization. This activity relies on the C-terminal microtubule binding basic domain and is regulated by phosphorylation by GSK3β.[50] In this case too, EB1 may play a key role since EB1 depletion abolishes microtubule stabilization.[46] Additionally, APC may regulate microtubule stabilization and dynamics through its interaction with the microtubule-destabilizing protein XMCAK.[62] Finally, APC promotes microtubule bundling[58] and this effect may be supported by its interaction with the scaffold protein PSD-95.[63] In conclusion, microtubule-associated APC directly or indirectly promotes microtubule elongation and stability. Consistent with this idea, APC forms clusters at the basal ends of microtubules in polarized epithelial cells and contributes to the formation of large bundles of parallel microtubules.[23] However, the contribution of such microtubule regulation to cell migration and more particularly to epithelial cell migration, is still unclear. Recent evidence indicates that loss of APC and consecutive perturbation of microtubule stability correlate with disappearance of cell protrusions and decreased epithelial cell migration.[64] It is probable that APC contributes its microtubule regulatory function to microtubule-dependent protrusions. Accordingly, APC has been shown to be involved in neurite outgrowth, a process that is dependent on microtubule elongation.[65] However, epithelial cell migration appears essentially actin dependent and microtubule-independent[66-68] pointing to a possible role of APC in actin regulation.

APC and Regulation of the Actin Cytoskeleton

APC translocation along microtubules to membrane protrusions leads to an increased concentration of APC and APC-associated proteins to the leading edge and locally promotes APC-mediated signalling pathways. Evidence for the signalling properties of APC includes its ability to bind and stimulate Asef, a Rac guanine exchange factor, through the N-terminal Armadillo repeat region of APC.[69] Truncated APC proteins like those expressed in colorectal tumour cells constitutively stimulate Asef activity. This leads to a decrease in E-cadherin-mediated adherens junctions and to an increase in spreading, membrane ruffling and migration of cultured epithelial cells.[70] Conversely, depletion of Asef in colorectal tumour cells inhibits cell migration. IQGAP1 provides another link between APC and Rho GTPases. Depletion of APC inhibits IQGAP1 accumulation at the leading edge and prevents both formation of the actin meshwork at the leading edge and cell migration.[32] IQGAP1 has recently been shown to stimulate actin assembly via the Arp2/3 complex[71] and it is tempting to speculate that APC recruits and perhaps activates IQGAP1 at the leading edge of migrating cells.

Because APC regulates microtubule and actin cytoskeletons by interacting with various partners, APC may be a key regulator of cell migration. This is supported by the fact that APC depletion in mice abolishes intestinal cell migration along the crypt-villus axis.[9] These apparent defects in cell migration may however result from APC's contribution to the canonical Wnt pathway that governs normal in vivo cell differentiation. In cultured cells, overexpression of APC promotes cell protrusions and loss of APC inhibits cell protrusions and migration, indicating that *Apc* mutations can affect cell migration independently of any effect on cell differentiation.[72] Interestingly, truncation in the APC protein (which results in the deficient binding of most cytoskeletal APC partners and in particular of those involved in the regulation of microtubule stability and cell polarity (Fig. 1)) is associated with disorganized, nondirected migration, leading to the idea that truncated APC is not totally inactive but still able to promote migration.[12-14] It also suggests that APC may contribute to direction sensing during epithelial cell migration.

APC and Cell Polarity

A migrating cell is characterized by the polarized organization of its cytoskeleton towards the direction of migration. For a cell to move in a given direction, such as the crypt-villus axis, cell polarity must be correctly oriented. While in most cell types cell migration per se is dependent on the actin cytoskeleton, directionality of migration is largely dependent on microtubules.[73] Direction sensing involves evolutionary conserved polarity proteins and leads to a polarized regulation of microtubule dynamics, associated with microtubule anchoring at the cell cortex. Although not fully demonstrated yet, the current hypothesis is that forces exerted on anchored microtubules promote centrosome reorientation and polarized membrane trafficking.[74]

APC, Microtubule Anchoring and Centrosome Reorientation

Because APC localizes at microtubule plus-ends, regulates microtubule dynamics and associates with microtubule plus-end proteins, APC has emerged as a key regulator of the cortical microtubule network and of migration directionality. In polarized epithelial cells, APC clusters align along microtubules and seem to promote microtubule attachment and stabilization at the basal cortex.[75] In migrating astrocytes, APC clusters are in very close proximity to the plasma membrane and depletion of APC perturbs microtubule interaction with the plasma membrane, suggesting that APC may be directly involved in microtubule anchoring.[36] Several binding partners could form a molecular bridge between microtubule-bound APC and the actin cytoskeleton or the plasma membrane (Fig. 3). β-catenin and IQGAP1 are both concentrated at the leading edge of migrating cells where they interact with both APC and the actin cytoskeleton.[29,30,32,76] Moreover, the final carboxy-terminal amino acids of APC form a PDZ binding domain that allows its binding to scaffolding proteins such as Dlg (SAP-97) or PSD-95 (SAP-90).[63,77] Colocalization of APC with Dlg has been observed at the tips of cellular protrusions in epithelial cells and astrocytes and

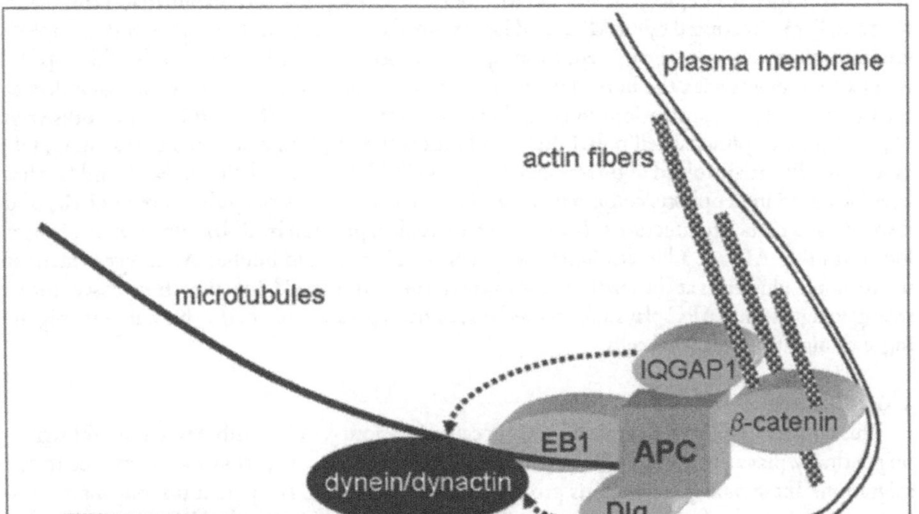

Figure 3. APC and its binding partners contribute to microtubule anchoring at the cell cortex. APC interacts with a large variety of cytoskeleton associated proteins (see Fig. 1). Here are depicted APC complexes shown to participate in microtubule capture and anchoring. APC directly interacts with several proteins (round shapes) that bind directly or indirectly (dotted arrows) with actin fibres or the microtubule-associated motor complex dynein/dynactin (in black). It is not known whether all these proteins are part of the same complex, or if they participate in several complexes containing different subsets of these proteins.

inhibition of the APC-Dlg interaction or depletion of Dlg strongly affects microtubule capture and microtubule network organization.[36,78] The formin mDia localizes at cortical microtubule ends where it may interact with APC and EB1, facilitate microtubule stabilization and participate to microtubule network polarization.[46]

APC may be involved in the interaction between microtubule plus-ends and the cell cortex and it may contribute to the recruitment and/or regulation of dynein, a microtubule associated motor involved in centrosome reorientation and directed migration[79,80] (Fig. 3). Similarly to dynein functional inhibition, APC depletion by RNA interference in astrocytes inhibits centrosome and Golgi reorientation and perturbs directed migration.[36] Via EB1, APC could be connected to dynein and to the dynein regulator dynactin.[81] However, the presence of APC in a complex with EB1, dynein and dynactin has not been demonstrated. Several other APC partners may also participate in this complex (Fig. 3). Via its interaction with IQGAP1, APC may regulate the plus-end associated protein CLIP170.[32] CLIP170 acts in concert with EB1 to control microtubule dynamics and stability[82,83] and it interacts with dynactin at microtubule plus-ends.[84] Finally, Dlg forms a complex with the protein GKAP, which binds dynein subunits[85] and depletion of Dlg abolishes centrosome reorientation and inhibits directed migration.[36] Taken together, APC interacts with a host of proteins at the leading edge and at the distal tips of microtubules and although APC clearly affects microtubule stability and dynamics, the precise mechanisms by which APC complexes regulate directional cell migration remain poorly understood.

APC and Polarity Proteins

Interestingly, evidence shows that APC can interact with several evolutionary conserved polarity proteins (Fig. 1). The best characterized role of APC in the regulation of polarity proteins comes from studies on developing neurons. Par3, the mammalian orthologue of the C. elegans PAR-3 protein,[86] controls axon positioning and growth during neuron development. Par3 has also been involved in epithelial cell polarity both in Drosophila and mammalian cells.[87,88] In neurons, Par3 is recruited by an APC- and kinesin-mediated transport to the plus-ends of rapidly growing microtubules at the nascent axon tip. This process is spatially regulated by GSK-3β.[35,89] APC also interacts with the human protein Scribble (Scrib) and the two proteins colocalize at the tip of membrane protrusions in epithelial cells.[90] Interestingly, Drosophila Scrib works with Dlg to regulate epithelial cell polarization and differentiation.[91] In mammalian cells, Scrib, Dlg as well as APC are involved in directional migration.[36,92,93] Downregulation of Scrib and Dlg has been observed in neoplastic colon mucosa and correlates with lack of epithelial cell polarity and disorganized tissue architecture.[94] Knockdown of Scrib expression by RNA interference disrupts localization of APC at adherens junctions in epithelial cells[90] and inhibits APC recruitment at microtubule plus-ends at the leading edge of migrating astrocytes.[92] Together these observations strongly suggest that APC plays a conserved role in the regulation of polarity both in non migrating and migrating epithelial cells.

Conclusion

This chapter highlights the role of APC in cell migration. APC contributes to the Wnt signalling pathway, plays a role in spindle assembly and chromosome segregation and is involved in cell migration. These multiple functions probably explain why APC is a potent tumour suppressor. One major challenge for future research will be to determine how exactly APC mutations affect its functions. In the case of cell migration, why and how APC truncations induce or modify epithelial cell migration remains unclear. APC plays essential roles in cell-cell adhesion, Rho GTPase family cell signaling, microtubule regulation and cell polarization. Does the loss of APC's carboxy-terminal domain as observed in human tumors perturb all of APC's functions? Are these functions together responsible for the critical role of APC in cell migration, or are some pathways more important than others? Another major unanswered question is how APC function in cell migration impacts on colon cancer development. How is cell migration affected by the loss of the second *Apc* allele? Is perturbation of migration critical during polyp formation? Is it an early and essential step during tumour development or does it contribute to tumour invasiveness? Despite many years of active investigation, the intricate mechanistic workings of APC proteins and the specific consequences of their disruption have not been fully revealed.

Acknowledgements

The author wishes to acknowledge the support of the Centre National de la Recherche Scientifique, the Pasteur Institute, the Agence Nationale pour la Recherche, the Fondation de France.

References

1. Powell SM, Zilz N, Beazer-Barclay Y et al. APC mutations occur early during colorectal tumorigenesis. Nature 1992; 359(6392):235-237.
2. Potten CS, Booth C, Pritchard DM. The intestinal epithelial stem cell: the mucosal governor. Int J Exp Pathol 1997; 78(4):219-243.
3. Smith KJ, Johnson KA, Bryan TM et al. The APC gene product in normal and tumor cells. Proc Natl Acad Sci USA 1993; 90(7):2846-2850.
4. Lamprecht SA, Lipkin M. Migrating colonic crypt epithelial cells: primary targets for transformation. Carcinogenesis 2002; 23(11):1777-1780.
5. Moss SF, Liu TC, Petrotos A et al. Inward growth of colonic adenomatous polyps. Gastroenterology 1996; 111(6):1425-1432.
6. He TC, Sparks AB, Rago C et al. Identification of c-MYC as target of the APC pathway. Science 1998; 281:1509-1512.
7. McDonald SA, Preston SL, Lovell MJ et al. Mechanisms of disease: from stem cells to colorectal cancer. Nat Clin Pract Gastroenterol Hepatol 2006; 3(5):267-274.
8. Pinto D, Clevers H. Wnt control stem cells and differentiation in the intestinal epithelium. Exp Cell Res 2006; 306(2):357-363.
9. Sansom OJ, Reed KR, Hayes AJ et al. Loss of Apc in vivo immediately perturbs Wnt signaling, differentiation and migration. Genes Dev 2004; 18(12):1385-1390.
10. Hoier EF, Mohler WA, Kim SK et al. The Caenorhabditis elegans APC-related gene apr-1 is required for epithelial cell migration and Hox gene expression. Genes Dev 2000; 14(7):874-886.
11. Fearnhead NS, Wilding JL, Bodmer WF. Genetics of colorectal cancer: hereditary aspects and overview of colorectal tumorigenesis. Br Med Bull 2002; 64:27-43.
12. Mahmoud NN, Bilinski RT, Churchill MR et al. Genotype-phenotype correlation in murine Apc mutation: differences in enterocyte migration and response to sulindac. Cancer Res 1999; 59(2):353-359.
13. Mahmoud NN, Boolbol SK, Bilinski RT et al. Apc gene mutation is associated with a dominant-negative effect upon intestinal cell migration. Cancer Res 1997; 57(22):5045-5050.
14. Su LK, Johnson KA, Smith KJ et al. Association between wild type and mutant APC gene products. Cancer Res 1993; 53(12):2728-2731.
15. Oshima M, Oshima H, Kitagawa K et al. Loss of Apc heterozygosity and abnormal tissue building in nascent intestinal polyps in mice carrying a truncated Apc gene. Proc Natl Acad Sci USA 1995; 92(10):4482-4486.
16. Oshima M, Oshima H, Kobayashi M et al. Evidence against dominant negative mechanisms of intestinal polyp formation by Apc gene mutations. Cancer Res 1995; 55(13):2719-2722.
17. Oshima H, Oshima M, Kobayashi M et al. Morphological and molecular processes of polyp formation in Apc(delta716) knockout mice. Cancer Res 1997; 57(9):1644-1649.
18. Nathke IS, Adams CL, Polakis P et al. The adenomatous polyposis coli tumor suppressor protein localizes to plasma membrane sites involved in active cell migration. J Cell Biol 1996; 134(1):165-179.
19. Langford KJ, Lee T, Askham JM et al. Adenomatous polyposis coli localization is both cell type and cell context dependent. Cell Motil Cytoskeleton 2006; 63(8):483-492.
20. Neufeld KL, White RL. Nuclear and cytoplasmic localizations of the adenomatous polyposis coli protein. Proc Natl Acad Sci USA 1997; 94(7):3034-3039.
21. Fagman H, Larsson F, Arvidsson Y et al. Nuclear accumulation of full-length and truncated adenomatous polyposis coli protein in tumor cells depends on proliferation. Oncogene 2003; 22(38):6013-6022.
22. Bienz M, Hamada F. Adenomatous polyposis coli proteins and cell adhesion. Curr Opin Cell Biol 2004; 16:528-235.
23. Mogensen MM, Tucker JB, Mackie JB et al. The adenomatous polyposis coli protein unambiguously localizes to microtubule plus ends and is involved in establishing parrallel arrays of microtubule bundles in highly polarized epithelial cells. J Cell Biol 2002; 157:1041-1048.
24. Yu X, Waltzer L, Bienz M. A new Drosophila APC homologue associated with adhesive zones of epithelial cells. Nat Cell Biol 1999; 1:144-151.

25. McCartney BM, Dierick HA et al. Drosophila APC2 is a cytoskeletally-associated protein that regulates wingless signaling in the embryonic epidermis. J Cell Biol 1999; 146:1303-1318.
26. Senda T, Miyashiro I, Matsumine A et al. The tumor suppressor protein APC colocalizes with beta-catenin in the colon epithelial cells. Biochem Biophys Res Commun 1996; 223(2):329-334.
27. Miyashiro I, Senda T, Matsumine A et al. Subcellular localization of the APC protein: immunoelectron microscopic study of the association of the APC protein with catenin. Oncogene 1995; 11:89-96.
28. Rubinfeld B, Souza B, Albert I et al. The APC protein and E-cadherin form similar but independent complexes with alpha-catenin, beta-catenin and plakoglobin. J Biol Chem 1995; 270(10):5549-5555.
29. Rosin-Arbesfeld R, Ihrke G, Bienz M. Actin-dependent membrane association of the APC tumour suppressor in polarized mammalian epithelial cells. EMBO J 2001; 20(21):5929-5939.
30. Townsley FM, Bienz M. Actin-dependent membrane association of a Drosophila epithelial APC protein and its effect on junctional Armadillo. Curr Biol 2000; 10(21):1339-1348.
31. McCartney BM, McEwen DG, Grevengoed E et al. Drosophila APC2 and Armadillo participate in tethering mitotic spindles to cortical actin. Nat Cell Biol 2001; 3(10):933-938.
32. Watanabe T, Wang S, Noritake J et al. Interaction with IQGAP1 links APC to Rac1, Cdc42 and actin filaments during cell polarization and migration. Dev Cell 2004; 7(6):871-883.
33. Etienne-Manneville S, Hall A. Cdc42 regulates GSK3 and adenomatous polyposis coli (APC) to control cell polarity. Nature 2003; 421:753-756.
34. Wen Y, Eng CH, Schmoranzer J et al. EB1 and APC bind to mDia to stabilize microtubules downstream of Rho and promote cell migration. Nat Cell Biol 2004; 6(9):820-830.
35. Shi SH, Cheng T, Jan LY et al. APC and GSK-3beta are involved in mPar3 targeting to the nascent axon and establishment of neuronal polarity. Curr Biol 2004; 14(22):2025-2032.
36. Etienne-Manneville S, Manneville JB, Nicholls S et al. Cdc42 and Par6-PKCζ regulate the spatially localized association of Dlg1 and APC to control cell polarization. J Cell Biol 2005; 170:895-901.
37. Mimori-Kiyosue Y, Shiina N, Tsukita S. The dynamic behavior of the APC-binding protein EB1 on the distal ends of microtubules. Curr Biol 2000; 10(14):865-868.
38. Smith KJ, Levy DB, Maupin P et al. Wild-type but not mutant APC associates with the microtubule cytoskeleton. Cancer Res 1994; 54(14):3672-3675.
39. Munemitsu S, Souza B, Muller O et al. The APC gene product associates with microtubules in vivo and promotes their assembly in vitro. Cancer Res 1994; 54(14):3676-3681.
40. Su LK, Burrel M, Hill DE et al. APC binds to the novel protein EB1. Cancer Res 1995; 55:2972-2977.
41. Askham JM, Moncur P, Markham AF et al. Regulation and function of the interaction between the APC tumour suppressor protein and EB1. Oncogene 2000; 19(15):1950-1958.
42. Barth AI, Siemers KA, Nelson WJ. Dissecting interactions between EB1, microtubules and APC in cortical clusters at the plasma membrane. J Cell Sci 2002; 115(Pt 8):1583-1590.
43. Jimbo T, Kawasaki Y, Koyama R et al. Identification of a link between the tumour suppressor APC and the kinesin superfamily. Nat Cell Biol 2002; 4(4):323-327.
44. Zhou FQ, Zhou J, Dedhar S et al. NGF-induced axon growth is mediated by localized inactivation of GSK-3beta and functions of the microtubule plus end binding protein APC. Neuron 2004; 42(6):897-912.
45. Cui H, Dong M, Sadhu DN et al. Suppression of kinesin expression disrupts adenomatous polyposis coli (APC) localization and affects beta-catenin turnover in young adult mouse colon (YAMC) epithelial cells. Exp Cell Res 2002; 280(1):12-23.
46. Wen Y. EB1 and APC bind to mDia to stabilize microtubules downstream of Rho and promote cell migration. Nat Cell Biol 2004; 6:820-830.
47. Rubinfeld B, Albert I, Porfiri E et al. Binding of GSK3beta to the APC-beta-catenin complex and regulation of complex assembly. Science 1996; 272:1023-1026.
48. Gartner A, Huang X, Hall A. Neuronal polarity is regulated by glycogen synthase kinase-3 (GSK-3beta) independently of Akt/PKB serine phosphorylation. J Cell Sci 2006; 119(Pt 19):3927-3934.
49. Sharma M, Leung L, Brocardo M et al. Membrane localization of adenomatous polyposis coli protein at cellular protrusions: targeting sequences and regulation by beta-catenin. J Biol Chem 2006; 281(25):17140-49.
50. Zumbrunn J, Kinoshita K, Hyman AA et al. Binding of the adenomatous polyposis coli protein to microtubules increases microtubule stability and is regulated by GSK3 beta phosphorylation. Curr Biol 2001; 11:44-49.
51. Ha NC, Tonozuka T, Stamos JL et al. Mechanism of phosphorylation-dependent binding of APC to beta-catenin and its role in beta-catenin degradation. Mol Cell 2004; 15(4):511-521.
52. Birchmeier W, Hulsken J, Behrens J. Adherens junction proteins in tumour progression. Cancer Surv 1995; 24:129-140.

53. McCartney BM, Price MH, Webb RLH et al. Testing hypotheses for the functions of APC family proteins using null and truncation alleles in Drosophila. Development 2006; 133(12):2407-2418.
54. Hulsken J, Behrens J, Birchmeier W. Tumor-suppressor gene products in cell contacts: the cadherin-APC-armadillo connection. Curr Opin Cell Biol 1994; 6(5):711-716.
55. Barth AI, Pollack AL, Altschuler Y et al. NH2-terminal deletion of beta-catenin results in stable colocalization of mutant beta-catenin with adenomatous polyposis coli protein and altered MDCK cell adhesion. J Cell Biol 1997; 136(3):693-706.
56. Klingelhöfer J, Troyanovsky RB, Laur OY et al. Exchange of catenins in cadherin-catenin complex. Oncogene 2003; 22(8):1181-1188.
57. Faux MC, Ross JL, Meeker C et al. Restoration of full-length adenomatous polyposis coli (APC) protein in a colon cancer cell line enhances cell adhesion. J Cell Sci 2004; 117(Pt 3):427-439.
58. Deka J, Kuhlmann J, Muller O. A domain within the tumor suppressor protein APC shows very similar biochemical properties as the microtubule-associated protein tau. Eur J Biochem 1998; 253(3):591-597.
59. Kita K, Wittmann T, Nathke IS et al. Adenomatous polyposis coli on microtubule plus ends in cell extensions can promote microtubule net growth with or without EB1. Mol Biol Cell 2006; 17(5):2331-2345.
60. Nakamura M, Zhou XZ, Lu KP. Critical role for the EB1 and APC interaction in the regulation of microtubule polymerization. Curr Biol 2001; 11:1062-1067.
61. Hayashi I, Wilde A, Mal TK et al. Structural basis for the activation of microtubule assembly by the EB1 and p150Glued complex. Mol Cell 2005; 19(4):449-460.
62. Banks JD, Heald R. Adenomatous polyposis coli associates with the microtubule-destabilizing protein XMCAK. Curr Biol 2004; 14(22):2033-2038.
63. Takamori N, Shimomura A, Senda T. Microtubule-bundling activity of APC is stimulated by interaction with PSD-95. Neurosci Lett 2006; 403(1-2):68-72.
64. Kroboth K, Newton IP, Kita K et al. Lack of adenomatous polyposis coli protein correlates with a decrease in cell migration and overall changes in microtubule stability. Mol Biol Cell 2006; 18(3):910-918.
65. Votin V, Nelson WJ, Barth AI. Neurite outgrowth involves adenomatous polyposis coli protein and beta-catenin. J Cell Sci 2005; 118(Pt 24):5699-5708.
66. Nakamura M, Mishima H, Nishida T et al. Requirement of microtubule assembly for initiation of EGF-stimulated corneal epithelial migration. Jpn J Ophthalmol 1991; 35(4):377-385.
67. Rosen EM, Meromsky L, Goldberg I et al. Studies on the mechanism of scatter factor. Effects of agents that modulate intracellular signal transduction, macromolecule synthesis and cytoskeleton assembly. J Cell Sci 1990; 96(Pt 4):639-649.
68. Gipson IK, Westcott MJ, Brooksby NG. Effects of cytochalasins B and D and colchicine on migration of the corneal epithelium. Invest Ophthalmol Vis Sci 1982; 22(5):633-642.
69. Kawasaki Y, Senda T, Ishidate T et al. Asef, a link between the tumor suppressor APC and G-protein signaling. Science 2000; 289(5482):1194-1197.
70. Kawasaki Y, Sato R, Akiyama T. Mutated APC and Asef are involved in the migration of colorectal tumour cells. Nat Cell Biol 2003; 5(3):211-215.
71. Le Clainche C, Schlaepfer D, Ferrari A et al. IQGAP1 stimulates actin assembly through the N-WASP-ARP2/3 pathway. J Biol Chem 2007; 282(1):426-435.
72. Kroboth K, Newton IP, Katsuhiro K et al. Lack of adenomatous polyposis coli protein correlates with a decrease in cell migration and overall changes in microtubule stability. Mol Biol Cell 2007; 18(3):910-918.
73. Etienne-Manneville S. Actin and microtubules in cell motility: which one is in control? Traffic 2004; 5:470-477.
74. Manneville JB, Etienne-Manneville S. Positioning centrosomes and spindle poles: looking at the periphery to find the centre. Biol Cell 2006; 98(9):557-565.
75. Reilein A, Nelson WJ. APC is a component of an organizing template for cortical microtubule networks. Nat Cell Biol 2005; 7(5):463-473.
76. Fukata M, Watanabe T, Noritake J et al. Rac1 and Cdc42 capture microtubules through IQGAP1 and CLIP-170. Cell 2002; 109(7):873-885.
77. Matsumine A, Ogai A, Senda T et al. Binding of APC to the human homolog of the Drosophila discs large tumor suppressor protein. Science 1996; 272(5264):1020-1023.
78. Iizuka-Kogo A, Shimomura A, Senda T. Colocalization of APC and DLG at the tip of cellular protrusions in cultured epithelial cells and its dependency on cytoskeletons. Histochem Cell Biol 2005; 123(1):67-73.
79. Etienne-Manneville S, Hall A. Integrin-mediated Cdc42 activation controls cell polarity in migrating astrocytes through PKCζ. Cell 2001; 106:489-498.

80. Palazzo AF, Cook TA, Alberts AS et al. mDia mediates Rho-regulated formation and orientation of stable microtubules. Nat Cell Biol 2001; 3:723-729.
81. Berrueta L, Tirnauer JS, Schuyler SC et al. The APC-associated protein EB1 associates with components of the dynactin complex and cytoplasmic dynein intermediate chain. Curr Biol 1999; 9(8):425-428.
82. Komarova Y, Lansbergen G, Galjart N et al. EB1 and EB3 control CLIP dissociation from the ends of growing microtubules. Mol Biol Cell 2005; 16(11):5334-5345.
83. Akhmanova A, Hoogenraad CC, Drabek K et al. CLASPs are Clip-115 and -170 associated proteins involved in the regional regulation of microtubule dynamics in motile fibroblasts. Cell 2001; 104:923-935.
84. Vaughan KT, Tynan SH, Faulkner NE et al. Colocalization of cytoplasmic dynein with dynactin and CLIP-170 at microtubule distal ends. J Cell Sci 1999; 112(Pt 10):1437-1447.
85. Haraguchi K, Satoh K, Yanai H et al. The hDLG-associated protein DAP interacts with dynein light chain and neuronal nitric oxide synthase. Genes Cells 2000; 5(11):905-911.
86. Izumi Y, Hirose T, Tamai Y et al. An atypical PKC directly associates and colocalizes at the epithelial tight junction with ASIP, a mammalian homologue of Caenorhabditis elegans polarity protein PAR-3. J Cell Biol 1998; 143(1):95-106.
87. Rolls MM, Albertson R, Shih HP et al. Drosophila aPKC regulates cell polarity and cell proliferation in neuroblasts and epithelia. J Cell Biol 2003; 163(5):1089-1098.
88. Ebnet K, Suzuki A, Horikoshi Y et al. The cell polarity protein ASIP/PAR-3 directly associates with junctional adhesion molecule (JAM). EMBO J 2001; 20:3738-3748.
89. Shi SH, Jan LY, Jan YN. Hippocampal neuronal polarity specified by spacially localized mPar3/mPar6 and PI3-Kinase activity. Cell 2003; 112:63-75.
90. Takizawa S, Nagasaka K, Nakagawa S et al. Human scribble, a novel tumor suppressor identified as a target of high-risk HPV E6 for ubiquitin-mediated degradation, interacts with adenomatous polyposis coli. Genes Cells 2006; 11(4):453-464.
91. Humbert P, Russell S, Richardson H. Dlg, Scribble and lgl in cell polarity, cell proliferation and cancer. Bioessays 2003; 25(6):542-553.
92. Osmani N, Vitale N, Borg JP et al. Scrib controls Cdc42 localization and activity to promote cell polarization during astrocyte migration. Curr Biol 2006; 16(24):2395-2405.
93. Dow LE, Kauffman JS, Caddy J et al. The tumour-suppressor Scribble dictates cell polarity during directed epithelial migration: regulation of Rho GTPase recruitment to the leading edge. Oncogene 2007; 26:2272-2282.
94. Gardiol D, Zacchi A, Petrera F et al. Human discs large and scrib are localized at the same regions in colon mucosa and changes in their expression patterns are correlated with loss of tissue architecture during malignant progression. Int J Cancer 2006; 119(6):1285-1290.

Chapter 4

The APC-EB1 Interaction

Ewan E. Morrison*

Abstract

Changes in cell proliferation seen in cancers initiated by *adenomatous polyposis coli* (*APC*) gene mutation are driven by the loss of an ability to negatively regulate the canonical WNT signalling pathway. However, mutant APC proteins also lack the ability to interact with a number of other ligands and it is possible that the loss of these interactions could contribute to the phenotype or to the development of colorectal tumours. One such association is with the microtubule plus-end binding protein EB1. Originally identified as an APC binding partner, EB1 is now known to be part of an evolutionarily conserved family of proteins involved in the regulation of microtubule dynamics and microtubule-dependent processes. Roles for the interaction between APC and EB1 have been identified in cellular functions as diverse as directed cell migration and mitosis, with potentially important implications for the behaviour of both normal epithelial cells and colorectal cancer cells. In this chapter our current understanding of the functional role of the APC-EB1 interaction will be reviewed.

Introduction

The identification of APC as a tumour suppressor protein in colorectal cancer led to a concerted research effort aimed at identifying functionally important binding partners. The distribution of the mutations found in the *APC* gene and the observation that many of these mutations led to the production of truncated APC proteins in tumour cells naturally led researchers to focus their interest on the C-terminal half of the protein, which harbors binding sites for a number of candidate interactors. An interaction with β-catenin was eventually identified as the core of the tumour suppressor function of APC. Nevertheless, interest in other ligands has continued since it was appreciated that the loss of other interactions might also contribute to the phenotype or to the development of colorectal tumours. One of the proteins identified by yeast-2-hybrid analysis using a C-terminal fragment of APC as bait was a novel small protein named EB1.[1]

EB1 Is an Evolutionarily Conserved Microtubule Plus-End Binding Protein

EB1 is a relatively small protein; at 30 kDa it is approximately one-tenth the size of wild-type APC. Several years after its original identification, EB1 was shown to be a microtubule-associated protein (MAP) that localised to a subset of microtubule plus ends in interphase and mitotic cells.[2,3] Confirmation that EB1 was a member of a new class of MAPs that specifically tracked the growing plus end of microtubules, now termed "+TIPs", came several years later through the observation of EB1-GFP fusion protein behaviour in living cells.[4] The initial characterisation of EB1 was performed in parallel with the developing understanding that APC was also a MAP that was intimately associated with microtubule plus ends.[5,6] This led to early suggestions that EB1 might

*Ewan E. Morrison—CRUK Clinical Centre at Leeds, Leeds Institute of Molecular Medicine, St James's University Hospital, Leeds LS9 7TF, UK. Email: e.e.morrison@leeds.ac.uk

APC Proteins, edited by Inke S. Näthke and Brooke M. McCartney. ©2009 Landes Bioscience and Springer Science+Business Media.

help target APC to these sites.[4,7] The subsequent characterisation of EB1 was rapid. It included the identification of two other members of the human EB protein family, EB3 and the more divergent RP1 (EB2)[8-11] and the realisation that EB1 was a member of a highly conserved family of proteins with members in organisms as diverse as yeast,[12] slime moulds[13] and plants.[14] In short, all eukaryotes appear to possess at least one member of the EB protein family. Unsurprisingly, a dramatic increase in the volume of published work on EB1 proteins has ensued, implicating them in a variety of cellular processes. A full analysis of current thinking on EB1 protein function, which is now suspected to be fundamentally important to +TIP function as a whole, lies outside the scope of this chapter. Fortunately, a number of recent reviews exist to guide the reading of those interested in this topic.[15-18] It is worth noting, however, that in the specific context of its effects upon microtubule dynamics, EB1 is thought to decrease microtubule catastrophe frequencies in vertebrate cells. It is therefore considered to be a promoter of microtubule growth and stability.[19] It also appears to be an obligate +TIP: that is, it localises to all of the growing microtubule plus-ends within a cell.

The EB1 Binding Region in APC

The initial identification of EB1 was achieved using a comparatively small fragment of the APC *C*-terminus, aa 2559 to 2843[1] (Fig. 1A). This region lies downstream of and is separable from the basic, direct microtubule-binding region in APC.[7,20-23] The precise location of the EB1 binding site was subsequently refined and shown to lie within the final 170aa of APC.[7] The organisation of the EB1 binding region in APC was recently subjected to detailed examinations that suggest a minimal EB1 binding site is encompassed by aa2782-2819, centred around aa2805-2806.[24,25] The interaction between the two proteins is regulated by APC phosphorylation,[7,26,27] potentially by CDK1 in mitotic cells at the conserved consensus sequence between Serine2789 and Lysine2792 (Fig. 1A).[7,26] However, conclusive identification of the kinase(s) that mediate such regulation in vivo (or indeed the phosphatases that presumably reverse this) has not yet been reported. The precise organisation of the APC-EB1 complex also remains to be fully defined. For example, although EB1 dimerisation appears to be crucial for this interaction,[25] it is unclear whether APC dimerisation is necessary.

In addition, the EB1 binding region in APC may not be fully defined. Yeast-2-hybrid and biochemical evidence has been presented showing that the human APC homolog APC2 (APC-L) interacts with EB family proteins.[10] Compared to APC, APC2 lacks much of the currently accepted EB1 binding domain. Instead, a different region with homology to APC aa2518-2577 was found to mediate binding (Fig. 1A), raising the possibility that another (or an extended) EB1 binding region exists in APC. However, a more recent study has suggested that this may not be the case and found no evidence for EB1 binding by the APC region with homology to the putative EB1 binding site in APC2.[28] This issue therefore awaits final clarification.

Finally, a recent study has suggested that in addition to mediating an association with microtubules, the basic domain in APC (Fig. 1A) may also mediate a direct association with the actin cytoskeleton. Intriguingly, this association was inhibited by binding of EB1 to APC, suggesting a novel regulatory function for the interaction.[29]

The APC Binding Site in EB1

The structural and functional organisation of EB1 proteins is now well understood (Fig. 1B). The *N*-terminal half of the EB1 molecule contains a calponin homology domain that mediates its interaction with microtubules[31,32] while the C-terminal half of the protein is responsible for both the interactions with all other ligands identified so far, and for EB1 dimerisation (Fig. 1B).[24,25,28,29,31,32] The interaction between EB1 and its different ligands is competitive.[31] Of the three EB family proteins in humans, EB1 and EB3 appear to interact well with APC whereas RP1 interacts much more poorly.[28]

Work on the APC-EB1 interaction provided the first evidence for the best current model of how the effect of EB1 on microtubule dynamics is regulated. It was noted that EB1 was only capable of

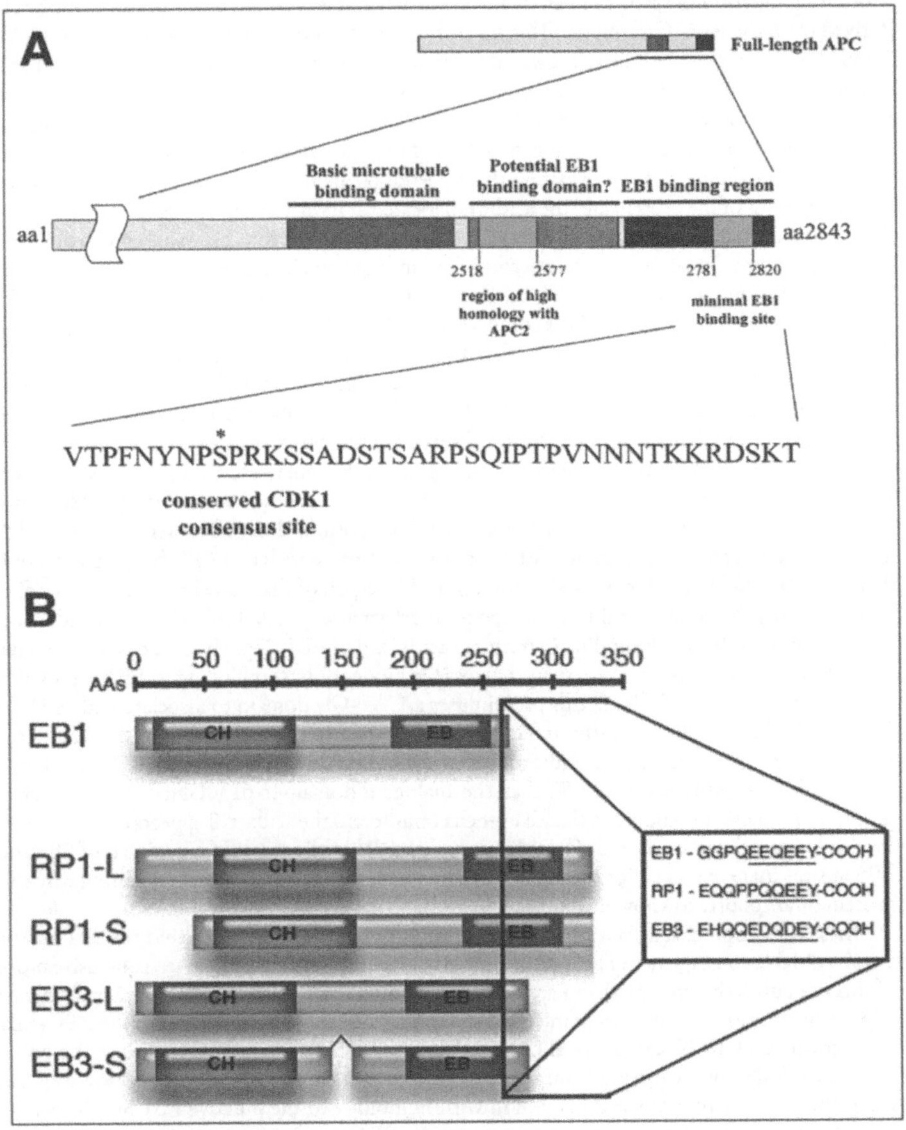

Figure 1. A) The EB1 binding site in APC. A detailed view of the EB1 binding region at the human APC C-terminus is shown, with the sequence of the core EB1 binding region containing a potential regulatory phosphorylation site highlighted. B) The human EB protein family. Three EB family (*MAPRE*) genes have been identified in humans. EB1 exists in one protein form. Two RP1 isoforms are generated by the use of alternative translation initiation sites and two EB3 isoforms are produced by alternative mRNA splicing. All family members share a common general structure with a microtubule-binding CH domain in the N-terminal half of the protein and a conserved ("EB") basic leucine zipper-like domain in the C-terminal half of the protein that mediates dimerisation and contributes to the binding of some ligands, including APC. An acidic C-terminal tail common to all protein forms of the human EB family is highlighted; this participates in the interaction with the CAP-Gly domains in p150Glued and CLIP-170 and the more divergent tail in RP1 may help to explain the lower affinity of this protein for most of the EB protein ligands examined to date.

promoting microtubule growth in an in vitro assay when it was bound to a recombinant protein derived from the APC C-terminus. This APC fragment did not contain an independent microtubule binding region and the effect was abrogated by APC phosphorylation, confirming that it required a direct interaction between the proteins.[26] A more recent study examining the interaction between EB1 and another of its ligands, the dynactin subunit p150Glued,[31] has provided an elegant explanation for this observation.[33] It is now thought that when not bound to ligand, EB1 exists in a closed, auto-inhibited conformation that depends on an intramolecular interaction between the N-terminus of the protein and the acidic tail located at its extreme C-terminus (Fig. 1B). This form of the protein cannot interact with microtubules or promote their growth. Binding of ligand opens the EB1 molecule, freeing the microtubule-binding domain, stabilising the *C*-terminal part of the protein and perhaps promoting the formation of a stable EB1 dimer.

Although their binding to EB1 is competitive, APC and other EB1 binding partners exploit different structural determinants in the EB1 C-terminus for their interaction with this protein. Structural studies indicate that EB1 dimerises via a basic leucine zipper-like domain located near the C-terminus of the protein, forming a 4-helix bundle that is stabilised by sequences C-terminal to the dimerisation domain. This results in a Y-shaped arrangement of globular domains visible by transmission electron microscopy, with a parallel pair of CH domains forming the arms of the Y.[25] EB1-APC binding involves the placement of a conserved isoleucine in human APC (Ile2805) into a hydrophobic cavity formed by EB1 C-terminal dimerisation, alongside other interactions.[24,25] The p150Glued interaction region in EB1 extensively overlaps with this APC binding site, but is demonstrably different since limited mutation in this region of EB1 can abrogate APC binding while leaving p150Glued binding unchanged.[30] Furthermore, p150Glued exhibits an additional interaction with the α-tubulin-like C-terminal acidic tail of EB1 (Fig. 1B) that has not been described for APC. Additional variety in the possible modes of ligand binding to EB1 is provided by CLIP-170. Like p150Glued, this protein uses a CAP-Gly domain to associate with EB1. In the case of CLIP-170 however, this interaction is almost entirely dependent upon binding to the EB1 acidic tail, with recent work indicating that the CAP-Gly domain in CLIP-170 has a much higher affinity for this feature of EB1 than the analogous domain in p150Glued.[34,35] In the latter protein, simultaneous binding to the EB1 4-helix bundle and the acidic tail appears to compensate for this lower affinity for the tail. The sequence of the RP1 acidic tail differs from that of EB1 and EB3 at sites that are crucial for the CLIP-170 interaction (Fig. 1B), perhaps explaining why this protein binds poorly to known EB protein ligands.

These data imply that activation of the microtubule growth promoting activity of EB1 may be an unavoidable consequence of ligand binding to the EB1 C-terminus. However, conclusive proof of this remains to be presented, in part because EB1 interacts promiscuously with a growing list of proteins that localise to or act at microtubule ends.[13-15] In the specific context of the APC-EB1 interaction, work by Nakamura et al (2001) clearly indicates that an APC fragment composed of the final 283aa of the protein is sufficient to bind to and activate EB1 in vitro.[26] However, this fragment contains part of the APC region with homology to the putative EB1 binding site in APC2 and may therefore encompass more than one EB1 interaction site. It is currently unclear whether smaller fragments of the APC C-terminus retain the ability to activate EB1, although they are clearly capable of binding to it.[7,24,25]

Interphase Functions of the APC-EB1 Interaction

Balancing all available evidence indicates that EB1 does not significantly contribute to the delivery of APC to cortical sites located near microtubule ends,[36,37] a role that is instead likely to be filled by an interaction between APC and kinesins.[6,38,39] In addition, EB1 does not appear to be necessary for the cortical clustering of APC at microtubule ends.[36,40] However, a functional role for an interaction between APC and EB1 at microtubule ends in the process of directed cell migration has been described.[30,41]

Many types of migrating cells orient their microtubule network to point their plus-ends towards the leading edge of the cell. A subset of microtubules in this array display markers indicative of stabilisation, a process that could facilitate sustained membrane delivery and the trafficking of signalling factors to the rapidly advancing cell front. This stabilisation occurs downstream of microtubule capture by cortically-located factors in conjunction with the "capping" of microtubule plus ends, a state in which microtubule plus end dynamics is suppressed.[17,42] The APC-EB1 interaction is thought to play a role in several microtubule capture and capping mechanisms at the cortex of migrating cells. Current models designate APC as the link to the cell cortex. This may be mediated by an interaction with the PDZ domain-containing cortical scaffolding protein DLG1[43] or through binding to factors that provide linkage to the cortical actin network. These include proteins such as the Rac1/Cdc42 effector protein IQGAP1[44] and the RhoA effector protein mDia.[30] mDia also interacts with EB1, suggesting that it may form part of a multiprotein complex with APC and EB1 that mediates microtubule cortical capture and stabilisation.[30] In both processes, the linkage to the microtubule tip is mediated by EB1, so the EB1-APC interaction provides the bridge between the microtubule end and the cell cortex. Notably, both of the above mechanisms are accessible to inputs from the Rho family small GTPase signalling pathways, which are known to play a crucial role in coordinating cell migration.[45]

One general caveat about the above studies is that they examine modes of migration more typical of fibroblasts than epithelial cells, the cell type in which APC mutation causes cancer. Recent work indicates that the role of APC and, by implication, the function of its interaction with EB1, might be different in the coordinated cell movements typical of the migration of epithelial monolayers.[46] Another potential problem lies in the recent discovery that ligand binding activates the microtubule growth promoting activity of EB1.[33] This suggests that APC binding to EB1 could drive microtubule polymerisation rather than stabilisation, although the latter could explain the anti-catastrophe activity of EB1. This conflict is likely the result of our current lack of understanding about how EB1 binds to microtubule plus-ends. Furthermore, recent careful studies suggest that cortically located APC on microtubule ends is capable of increasing microtubule stability and driving microtubule growth independently of EB1.[37] This may represent a difference between two independent processes: the promotion of microtubule growth by APC in dynamic clusters at microtubule ends and the coordinated capture and capping of microtubules in migrating cells.

An additional role for the APC-EB1 interaction in interphase has been proposed that relates to the observation that both proteins localise to centrosomes.[2,3,41] A detailed analysis indicated that both specifically localised to the sub-distal appendages of the mother centriole in a microtubule-independent manner.[47] The finding that EB1 is involved in anchoring of microtubules at centrosomes [31] raises the possibility that an interaction between APC and EB1 at this site might contribute to this process. However, a number of other EB1 ligands also localise to centrosomes and play a functional role in microtubule anchoring,[18,48] so the delineation of a specific function for a centrosomal APC-EB1 interaction will have to be carefully dissected.

Finally, evidence for a further, unexpected consequence of EB1 binding to APC has recently been presented.[49] EB1 overexpression in transfected cells was found to induce the nuclear accumulation of β-catenin and promote the transcriptional activity of TCF, potentially by directly antagonising the interaction between APC and β-catenin. This raises the interesting possibility that overexpressed EB1 might act as an oncogenic factor by deregulating WNT signaling. Supporting this idea, EB1 overexpression in human esophageal squamous cell carcinoma correlated with increased levels of cytoplasmic and nuclear β-catenin.[49] Since EB1 is small in comparison to APC and the binding sites for EB1 and β-catenin are well separated, this intriguing observation also implies that when APC binds EB1 the conformation of the former is changed in a way that inhibits its association with β-catenin.

Mitotic Functions of the APC-EB1 Interaction

Chromosomal instability (CIN) is a common feature of colorectal cancers initiated by *APC* mutation.[50,51] The possibility that this form of genomic instability is directly related to the loss of an APC function separate from its role in the WNT signalling pathway is seductive, since it implies that APC is a tumour suppressor protein with integral "gatekeeper" and "caretaker" functions in the intestinal epithelium.[52] If true, this would mean that loss of APC function drives both cellular hyper-proliferation and the increased mutation rate needed for the progression from adenoma to carcinoma. Since CIN could arise from defective chromosomal segregation during mitosis, which is a microtubule-based process, the potential roles of APC and EB1 in the mitotic spindle have received much attention.

The first detailed examinations of this issue in mammalian systems focused on APC, to yield results consistent with a role for the protein in ensuring the fidelity of chromosomal segregation.[53,54] However, a major hurdle for the concept that APC mutation could be a cause of CIN in colorectal tumour cells came in the form of the spindle assembly checkpoint. This process monitors both microtubule attachment to kinetochores and chromosomal alignment during metaphase, preventing entry into anaphase if abnormalities occur. Put simply, if APC acted in concert with EB1 to ensure normal chromosomal congression, defects in this process due to APC mutation should prevent affected cells from completing mitosis unless the spindle assembly checkpoint was also compromised.[55] However, in general, colorectal cancer cells appear to possess a functional checkpoint mechanism,[56] although evidence that mutant APC proteins might induce defects in this monitoring process has also been presented.[57]

Recent studies have directly addressed this issue in detail and have provided a viable explanation for how this problem might be circumvented in tumour cells.[58-60] Knockdown of APC and EB1 expression using inhibitory RNA technology confirmed that neither protein was essential for chromosomal congression. However, knock down of either protein induced an unusual phenotype consisting of chromosomes that were sufficiently well congressed to meet the stringent requirements of the spindle assembly checkpoint but more disordered in their alignment within the metaphase plate than in a normal mitosis. This disorder, or a further consequence of the defective process that allowed it, led to a low but detectable level of aberrant chromosomal segregation during anaphase and sporadic chromosome loss.[60] Expression of a dominant-negative mutant EB1 protein produced a similar phenotype.[60] Furthermore, expression of a truncated APC protein was observed to have a dominant-negative effect on the interaction between wild type APC and EB1 and on chromosome segregation [59] (this is discussed in more detail in the chapters by Caldwell and Kaplan this volume) suggesting that mutation of the first copy of *APC* in a colorectal epithelial cell might directly increase the possibility of losing the second copy via CIN, thus greatly increasing the odds of a carcinoma developing.

The specific metaphase process that might be compromised by loss of the APC-EB1 interaction in mitotic cells remains unclear. Since the defect affects the dynamics and orientation of sister chromatids in the congressed metaphase plate, but not to an extent that can be sensed by the spindle assembly checkpoint, it is likely to be subtle. Given the effects of EB1 on microtubule dynamics one obvious possibility is that the behaviour of microtubule plus ends attached to the kinetochore is perturbed, perhaps because activation of EB1-dependent microtubule growth by kinetochore-associated APC is lost. However, other potential mechanisms also exist. For example, recent work has indicated that the inhibition of APC expression compromises the spindle assembly checkpoint itself, reducing the kinetochore association of core checkpoint components.[61] Furthermore, the possibility that APC phosphorylation could block the interaction with EB1 in mitotic cells has not been addressed in the context of this work. Nevertheless, regardless of the specific mechanisms and regulatory pathways involved, the hypothesis that *APC* mutation could directly contribute to aneuploidy in colorectal cancer cells remains an important topic for closer study.

Another effect observed in APC and EB1 knockdown experiments[59,60] but also in a more recent study focusing on the role of EB1 in spindle orientation,[62] was defective spindle positioning in mitotic cells. This phenotype could be linked to defects in astral microtubule nucleation, anchoring or function. In epithelial cells, spindle positioning is an active process in which astral microtubule-dependent mechanisms specify the orientation of the mitotic spindle and hence the plane of cell division.[63-65] Defects in this process could therefore lead to a progressive loss of normal tissue organisation in polyps and tumours initiated by *APC* mutation.

APC-EB1 Interactions in Other Organisms

Although parallels have been drawn between the potential function of an APC-EB1 interaction in vertebrate cells and processes such as spindle movement in budding yeast,[17,66-68] it seems likely that APC has evolved to play a developmental role in complex, multi-cellular animals. As such, *Drosophila* is the most genetically tractable system in which its function has been examined to date. Two APC proteins have been identified in fruit flies, APC1 and APC2.[69,70-72]

APC1 appears to be the closest orthologue of human APC, with a high degree of domain conservation and a largely microtubule-dependent localisation, whereas APC2 is more divergent in both sequence and localisation and is primarily located at the cortex of cells. Unfortunately, neither *Drosophila* APC protein possesses a C-terminal region similar to the core site shown to interact with EB1 in vertebrate systems[24,25] and no evidence for a physical interaction between fly APC1 or APC2 and EB1 has been reported. Nevertheless, roles for both APC proteins in spindle orientation have been presented[73-75] and an overlapping role for EB1 has also been observed.[73] However, more recent work indicates that the fly APC proteins may not play a major role in spindle formation or positioning,[76] suggesting that further clarification of this issue may be necessary. In contrast, *Drosophila* EB1 clearly plays an important role during mitosis where it is required for both the assembly and the correct positioning of the spindle.[77] On the basis of the evidence presented to date, it seems reasonable to assume that if APC and EB1 proteins have mitotic roles in *Drosophila* cells, they perform them either indirectly in the same pathway or in parallel in separate pathways rather than through a direct interaction. Interestingly, this implies that the an APC-EB1 interaction is a comparatively recent event in evolution, perhaps occurring during the emergence of vertebrates.[24] Furthermore, it suggests that a clear definition of the role of the APC-EB1 interaction can only be found by work in vertebrate systems.

Conclusion

A decade of research has revealed important roles for the APC-EB1 interaction in a number of important processes, all of which could contribute to the phenotype of colorectal tumour cells when the interaction is lost following *APC* mutation. Loss of a function in directed cell migration could underpin the aberrant migration of cells along the crypt-villus axis, whereas subtle defects in chromosomal segregation might contribute to a CIN phenotype in colon cancers. In addition, an inability to correctly orient the mitotic spindle parallel with the epithelial cell monolayer lining the colon could contribute to a lack of normal tissue organisation in tumours. Finally, the possibility that EB1 overexpression could contribute to deregulated WNT signalling in other malignancies by negatively regulating APC function requires closer attention. However, none of the above functions have been completely defined and research into both APC and EB1 has demonstrated an unusual capacity to surprise and a tendency to reveal increasing complexity. An assumption that we now have a comprehensive understanding of either the normal APC-EB1 interaction or the implications of its loss in colorectal cancer cells is therefore premature.

Acknowledgments

The author wishes to acknowledge the past and present support of Yorkshire Cancer Research, the Medical Research Council (UK) and Cancer Research UK.

References

1. Su LK, Burrell M, Hill DE et al. APC binds to the novel protein EB1. Cancer Res 1995; 55:2972-2977.
2. Morrison EE, Wardleworth B, Askham J et al. EB1, a protein which interacts with the APC tumour suppressor, is associated with the microtubule cytoskeleton throughout the cell cycle. Oncogene 1998; 17:3471-3477.
3. Berrueta L, Kraeft SK, Tirnauer JS et al. The adenomatous polyposis coli-binding protein EB1 is associated with cytoplasmic and spindle microtubules. Proc Natl Acad Sci USA 1998; 95:10596-10601.
4. Mimori-Kiyosue Y, Shiina N, Tsukita S. The dynamic behaviour of the APC-binding protein EB1 on the distal ends of microtubules. Curr Biol 2000; 10:865-868.

5. Näthke IS, Adams CL, Polakis P et al. The adenomatous polyposis coli tumor suppressor protein localises to plasma membrane sites involved in active cell migration. J Cell Biol 1996; 134:165-179.
6. Mimori-Kiyosue Y, Shiina N, Tsukita S. APC protein moves along microtubules and concentrates at their growing ends in epithelial cells. J Cell Biol 2000; 148;505-517.
7. Askham JM, Moncur PM, Markham AF et al. Regulation and function of the interaction between the APC tumour suppressor protein and EB1. Oncogene 2000; 19:1950-1958.
8. Renner C, Pfitzenmeier JP, Gerlach K et al. RP1, a new member of the adenomatous polyposis coli-binding EB1-like gene family, is differentially expressed in activated T-cells. J Immunol 1997; 159:1276-1283.
9. Juwana JP, Henderikx P, Mischo A et al. EB/RP gene family encodes tubulin binding proteins. Intl J Cancer 1999; 81:275-284.
10. Nakagawa H, Koyama K, Murata Y et al. EB3, a novel member of the EB1 family preferentially expressed in the central nervous system, binds to a CNS-specific APC homologue. Oncogene 2000; 19:210-216.
11. Su LK, Qi Y. Characterization of human MAPRE genes and their proteins. Genomics 2001; 71:142-149.
12. Schwartz K, Richards K, Botstein D. BIM1 encodes a microtubule-binding protein in yeast. Mol Biol Cell 1997; 8:2677-2691.
13. Rehberg M, Graf R. Dictyostelium EB1 is a genuine centrosomal component required for proper spindle formation. Mol Biol 2002; 13:2301-2310.
14. Mathur J, Mathur N, Kernebeck B et al. A novel localization pattern for an EB1-like protein links microtubule dynamics to endomembrane organization. Curr Biol 2003; 13:1991-1997.
15. Tirnauer JS, Bierer BE. EB1 proteins regulate microtubule dynamics, cell polarity and chromosome stability. J Cell Biol 2000; 149:761-766.
16. Akhmanova A, Hoogenraad CC. Microtubule plus end tracking proteins: mechanisms and functions. Curr Opin Cell Biol 2005; 17:47-54.
17. Lansbergen G, Akhmanova A. Microtubule plus end: a hub of cellular activities. Traffic 2006; 7:499-507.
18. Morrison EE. Action and interactions at microtubule ends. Cell Mol Life Sci 2006; 64:307-317.
19. Tirnauer JS, Grego S, Salmon ED et al. EB1-microtubule interactions in xenopus egg extracts: role of EB1 in microtubule stabilization and mechanisms of targeting to microtubules. Mol Biol Cell 2002; 13:3614-3626.
20. Munemitsu S, Souza B, Muller O et al. The APC gene product associates with microtubules in vivo and promotes their assembly in vitro. Cancer Res 1994; 54:3676-3681.
21. Smith KJ, Levy DB, Maupin P et al. Wild-type but not mutant APC associates with the microtubule cytoskeleton. Cancer Res 1994; 54:3672-3675.
22. Deka J, Kuhlmann J, Muller O. A domain within the tumour suppressor protein APC shows very similar biochemical properties as the microtubule-associated protein tau. Eur J Biochem 1998; 253:591-597.
23. Zumbrunn J, Kinoshita K, Hyman AA et al. Binding of the adenomatous polyposis coli protein to microtubules increases microtubule stability and is regulated by GSK3β phosphorylation. Curr Biol 2001; 11:44-49.
24. Slep KC, Rogers SL, Elliott SL et al. Structural determinants for EB1-mediated recruitment of APC and spectraplakins to the microtubule plus end. J Cell Biol 2005; 168:587-598.
25. Honnappa S, John CM, Kostrewa D et al. Structural insights into the EB1-APC interaction. EMBO J 2005; 24:261-269.
26. Nakamura M, Zhou XZ, Lu KP. Critical role for the EB1 and APC interaction in the regulation of microtubule polymerization. Curr Biol 2001; 11:1062-1067.
27. Trzepacz C, Lowy AM, Kordich JJ et al. Phosphorylation of the tumor suppressor adenomatous polyposis coli (APC) by the cyclin-dependent kinase p34. J Biol Chem 1997; 272:21681-21684.
28. Bu W, Su LK. Characterization of functional domains of human EB1 family proteins. J Biol Chem 2003; 278:49721-49731.
29. Moseley JB, Bartolini F, Okada K et al. Regulated binding of adenomatous polyposis coli protein to actin. J Biol Chem 2007; 282:12661-12668.
30. Wen Y, Eng CH, Schmoranzer J et al. EB1 and APC bind to mdia to stabilize microtubules downstream of rho and promote cell migration. Nat Cell Biol 2004; 6:820-830.
31. Askham JM, Vaughan KT, Goodson HV et al. Evidence that an interaction between EB1 and p150(Glued) is required for the formation and maintenance of a radial microtubule array anchored at the centrosome. Mol Biol Cell 2002; 13:3627-3645.
32. Hayashi I, Ikura M. Crystal structure of the amino-terminal microtubule-binding domain of end-binding protein 1 (EB1). J Biol Chem 2003; 278:36430-36434.

33. Hayashi I, Wilde A, Mal TK et al. Structural basis for the activation of microtubule assembly by the EB1 and p150(Glued) complex. Mol Cell 2005; 19:449-460.
34. Komarova Y, Lansbergen G, Galjart N et al. EB1 and EB3 control CLIP dissociation from the ends of growing microtubules. Mol Biol Cell 2005; 16:5334-5345.
35. Mishima M, Maesaki R, Kasa M et al. Structural basis for tubulin recognition by cytoplasmic linker protein 170 and its autoinhibition. Proc Natl Acad Sci USA 2007; 104:10346-10351.
36. Barth AI, Siemers KA, Nelson WJ. Dissecting interactions between EB1, microtubules and APC in cortical clusters at the plasma membrane. J Cell Sci 2002; 115:1583-1590.
37. Kita K, Wittman T, Nathke IS et al. Adenomatous polyposis coli on microtubule plus ends in cell extensions can promote microtubule net growth with or without EB1. Mol Biol Cell 2006; 17:2331-2345.
38. Jimbo T, Kawasaki Y, Koyama R et al. Identification of a link between the tumour suppressor APC and the kinesin superfamily. Nat Cell Biol 2002; 4:323-327.
39. Langford KJ, Askham JM, Lee T et al. Examination of actin and microtubule dependent APC localisations in living mammalian cells. BMC Cell Biol 2006; 7:3.
40. Sharma M, Leung L, Brocardo M et al. Membrane localisation of adenomatous polyposis coli protein at cellular protrusions: targeting sequences and regulation by beta-catenin. J Biol Chem 2006; 281:17140-17149.
41. Etienne-Manneville S, Hall A. Cdc42 regulates GSK-3β and adenomatous polyposis coli to control cell polarity. Nature 2003; 421:753-756.
42. Mannevelle JB, Etienne-Manneville S. Positioning centrosomes and spindle poles: looking at the periphery to find the centre. Biol Cell 2006; 98:557-565.
43. Etienne-Manneville S, Manneville JB, Nicholls S et al. Cdc42 and Par6-PKCzeta regulate the spatially localised association of Dlgl and APC to control cell polarization. J Cell Biol 2005; 170:895-901.
44. Watanabe T, Wang S, Noritake J et al. Interaction with IQGAP links APC to Rac1, Cdc42 and actin filaments during cell polarization and migration. Dev Cell 2004; 7:871-883.
45. Watanabe T, Noritake J, Kaibuchi K. Regulation of microtubules in cell migration. Trends Cell Biol 2005; 15:76-83.
46. Langford KJ, Lee T, Askham JM et al. Adenomatous polyposis coli localization is both cell type and cell context dependent. Cell Motil Cytoskeleton 2006; 63:483-492.
47. Louie RK, Bahmanyar S, Siemers KA et al. Adenomatous polyposis coli and EB1 localize in close proximity of the mother centriole and EB1 is a functional component of centrosomes. J Cell Sci 2004; 117:1117-1128.
48. Yan X, Habedanck R, Nigg EA. A complex of two centrosomal proteins, CAP350 and FOP, cooperates with EB1 in microtubule anchoring. Mol Biol Cell 2006; 17:634-644.
49. Wang Y, Zhou X, Zhu H et al. Overexpression of EB1 in human esophageal squamous cell carcinoma (ESCC) may promote cellular growth by activating β-catenin/TCF pathway. Oncogene 2005; 24:6637-6645.
50. Lengauer C, Kinzler KW, Vogelstein B. Genetic instability in colorectal cancers. Nature 1997; 386:623-627.
51. Fodde R, Smits R, Clevers H. APC, signal transduction and genetic instability in colorectal cancer. Nat Rev Cancer 2001; 1:55-67.
52. Kinzler KW, Vogelstein B. Lessons from hereditary colorectal cancer. Cell 1996; 87:159-170.
53. Fodde R, Kuipers J, Rosenberg C et al. Mutations in the APC tumour suppressor gene cause chromosomal instability. Nat Cell Biol 2001; 3:433-438.
54. Kaplan KB, Burds AA, Swedlow J et al. A role for the adenomatous polyposis coli protein in chromosome segregation. Nat Cell Biol 2001; 3:429-432.
55. Draviam VM, Xie S, Sorger PK. Chromosomal segregation and genomic stability. Curr Op Genet Dev 2004; 14:120-125.
56. Tighe A, Johnson VL, Albertella M et al. Aneuploid colon cancer cells have a robust spindle checkpoint. EMBO Rep 2001; 2:609-614.
57. Tighe A, Johnson VL, Taylor SS. Truncating APC mutations have dominant effects on proliferation, spindle checkpoint control, survival and chromosomal stability. J Cell Sci 2004; 117:6339-6353.
58. Green RA, Kaplan KB. Chromosomal instability in colorectal tumor cells is associated with defects in microtubule plus-end attachments caused by a dominant mutation in APC. J Cell Biol 2003; 163:949-961.
59. Green RA, Wollman R, Kaplan KB. APC and EB1 function together in mitosis to regulate spindle dynamics and chromosome alignment. Mol Biol Cell 2005; 16:4609-4622.
60. Draviam VM, Shapiro I, Aldridge B et al. Misorientation and reduced stretching of aligned sister kinetochores promote chromosome missegregation in EB1- or APC-depleted cells. EMBO J 2006; 25:2814-2827.

61. Dikovskaya D, Schiffmann D, Newton IP et al. Loss of APC induces polyploidy as a result of a combination of defects in mitosis and apoptosis. J Cell Biol 2007; 176:183-195.
62. Toyoshima F, Nishida E. Integrin-mediated adhesion orients the spindle parallel to the substratum in an EB1- and myosin X-dependent manner. EMBO J 2007; 26:1487-1498.
63. Reinsch S, Karsenti E. Orientation of spindle axis and distribution of plasma membrane proteins during cell division in polarised MDCK-II cells. J Cell Biol 1994; 126:1509-1526.
64. Busson S, Dujardin D, Moreau A et al. Dynein and dynactin are localised to astral microtubules and at cortical sites in mitotic epithelial cells. Curr Biol 1998; 8:541-544.
65. Morrison EE, Askham JM. EB1 immunofluorescence reveals an increase in growing astral microtubule length and number during anaphase in NRK-52E cells. Eur J Cell Biol 2001; 80:749-753.
66. Muhua L, Adames NR, Murphy MD et al. A cytokinesis checkpoint requiring the yeast homolog of an APC-binding protein. Nature 1998; 393:487-491.
67. Korinek WS, Copeland MJ, Chaudhuri A et al. Molecular linkage underlying microtubule orientation towards cortical sites in yeast. Science 2000; 287:2257-2259.
68. Lee L, Tirnauer JS, Li J et al. Positioning of the mitotic spindle by a cortical-microtubule capture mechanism. Science 2000; 287:2260-2262.
69. Bienz M, Hamada F. Adenomatous polyposis coli proteins and cell adhesion. Curr Op Cell Biol 2004; 16:528-535.
70. Hayashi S, Rubinfeld B, Souza B et al. A Drosophila homolog of the tumor suppressor gene adenomatous polyposis coli down-regulates beta-catenin but its zygotic expression is not essential for the regulation of armadillo. Proc Natl Acad Sci USA 1997; 94:242-247.
71. McCartney BM, Dierick HA, Kirkpatrick C et al. Drosophila APC2 is a cytoskeletally-associated protein that regulates wingless signalling in the embryonic epidermis. J Cell Biol 1999; 146:1303-1318.
72. Yu X, Waltzer L, Bienz M. A new Drosophila APC homologue associated with adhesive zones of epithelial cells. Nat Cell Biol 1999; 1:144-151.
73. Lu B, Roegiers F, Jan LY et al. Adherens junctions inhibit asymmetric division in the drosophila epithelium. Nature 2001; 409:522-525.
74. McCartney BM, McEwen DG, Grevengoed E et al. Drosophila APC2 and aramadillo participate in tethering mitotic spindles to cortical actin. Nat Cell Biol 2001; 3:933-938.
75. Yamashita YM, Jones DL, Fuller M. Orientation of asymmetric cell division by the APC tumour suppressor and centrosome. Science 2003; 301:1547-1550.
76. McCartney BM, Price MH, Webb RL et al. Testing hypotheses for the function of APC family proteins using null and truncation alleles in drosophila. Development 2006; 133:2407-2418.
77. Rogers SL, Rogers CC, Sharp DJ et al. Drosophila EB1 is important for proper assembly, dynamics and positioning of the mitotic spindle. J Cell Biol 2002; 158:873-884.

Chapter 5

The Role of APC in Mitosis and in Chromosome Instability

Christine M. Caldwell and Kenneth B. Kaplan*

Abstract

The established role of APC in regulating microtubules and actin in polarized epithelia naturally raises the possibility that APC similarly influences the mitotic cytoskeleton. The recent accumulation of experimental evidence in mitotic cells supports this supposition. APC associates with mitotic spindle microtubules, most notably at the plus-ends of microtubules that interact with kinetochores. Genetic experiments implicate APC in the regulation of spindle microtubule dynamics, probably through its interaction with the microtubule plus-end binding protein, EB1. Moreover, functional data show that APC modulates kinetochore-microtubule attachments and is required for the spindle checkpoint to detect transiently misaligned chromosomes. Together this evidence points to a role for APC in maintaining mitotic fidelity. Such a role is particularly significant when considered in the context of the chromosome instability observed in colorectal tumors bearing mutations in *APC*. The prevalence of APC truncation mutants in colorectal tumors and the ability of these alleles to act dominantly to inhibit the mitotic spindle place chromosome instability at the earliest stage of colorectal cancer progression (i.e., prior to deregulation of β-catenin). This may contribute to the autosomal dominant predisposition of patients with familial adenomatous polyposis to develop colon cancer. In this chapter, we will review the literature linking APC to regulation of mitotic fidelity and discuss the implications for dividing epithelial cells in the intestine.

Introduction

Although our understanding of the range of activities associated with the protein encoded by the *adenomatous polyposis coli* gene (*APC*) is incomplete, an abundance of evidence suggests that APC acts to link signaling pathways with cell differentiation and cell division. Studied intensely in tumor cells and as a downstream effector of Wnt signaling (see Kennell and Cadigan, this volume), APC directly modulates the dual roles of β-catenin in cell adhesion and gene transcription. Perhaps more significantly for this discussion, APC acts to increase microtubule plus-end dynamics and may contribute to the ability of microtubules to form stabilizing plus-end attachments. A number of studies have linked APC and microtubules to the establishment of cell polarity;[1-3] for the purposes of this discussion, we consider cell polarity to generally include the establishment of a polarized cytoskeleton necessary for the proper distribution of proteins and organelles. The same machinery involved in establishing cell polarity is also likely to be critical for cell division, especially (but not limited to) in the context of highly organized epithelial monolayers. As cells in a monolayer enter mitosis, the cytoskeleton must be reorganized to allow the assembly of a bipolar mitotic spindle. Changes in microtubule dynamics lead to the capture of chromosomes via spindle

*Corresponding Author: Kenneth B. Kaplan—Section of Molecular and Cellular Biology, University of California, Davis, Davis CA 95616, USA. Email: kbkaplan@ucdavis.edu

APC Proteins, edited by Inke S. Näthke and Brooke M. McCartney. ©2009 Landes Bioscience and Springer Science+Business Media.

microtubules and kinetochores, ultimately aligning chromosomes in metaphase. This process is monitored by quality control components of the spindle checkpoint to ensure that chromosomes are properly segregated. Finally, spindle microtubules are critical for determining the orientation of the cleavage plane during cytokinesis and therefore may also impact the position and fate of cells in epithelial tissues such as in the intestine. Thus, the same elements of cell polarity affected by APC in migrating cells are also central to the success of mitosis. The goal of this review is to bridge our understanding of APC in polarized epithelial cells with its roles during mitosis. In particular, we will relate these roles to the normal division of cells in the intestinal crypt and how clinically relevant mutations in *APC* that affect mitosis may contribute to colorectal cancer (see summary in Fig. 3).

APC in Mitotic Cells

Progress in understanding the function of APC in migrating cells (see Etienne-Manneville, this volume) has provided important insights into the potential role of APC in mitosis. For example, the association of APC with microtubules and with the microtubule plus-end binding protein EB1 (see Morrison, this volume) focused attention on the ability of APC to regulate microtubule dynamics.[4,5] More recently, APC together with EB1, has been implicated in controlling the stability of microtubules in migrating cells downstream of the actin regulatory proteins, RhoA and the formin, mDia.[6,7] Before discussing what is known of APC in mitosis, it is useful to consider how the well-studied events in migrating cells relate to events in mitotic cells. Cell migration is accompanied by an increase in microtubule dynamics near the leading edge and at least some of these microtubules form stabilizing plus-end attachments at adhesive contacts (for review, see ref. 8). Similarly, the increase in microtubule dynamics associated with the onset of mitosis aids in the capture of microtubule plus-ends by kinetochores and their stabilization as kinetochore-microtubules (for review, see refs. 9 and 10). Thus, the ability to transform microtubule plus-ends from dynamic polymers to stable attachments may represent a function of APC that is shared between migrating and mitotic cells (see Etienne-Manneville, this volume).

Analysis of mitotic cells identified multiple populations of APC associated with the mitotic apparatus. In mouse embryonic stem cells, APC is found associated with kinetochores in metaphase and with polar microtubules in anaphase.[11] Similarly, in HeLa (human) cells and PtK (rat kangaroo) cells, APC is found associated with outer kinetochores, distinct from inner kinetochore complexes such as CENP A.[12] APC is also observed on centrosomes (spindle poles) and generally throughout the cytoplasm, consistent with the presence of multiple pools of APC in mitotic cells. Unlike resident kinetochore proteins, depolymerization of microtubules prevents APC from localizing to kinetochores, arguing that APC is associated with microtubule plus-ends at the kinetochore.[11,12] Indeed, high resolution microscopy revealed APC along the length of growing kinetochore microtubules, raising the possibility that APC uses microtubules to traffic to specific subcellular sites. Interestingly, the carboxy terminal domains of APC that interact with EB1 and with microtubules are not required for its mitotic spindle localization,[13] although they are required for APC to stabilize microtubules in vitro.[14] Thus, both wild type and mutant APC that lacks the carboxy terminus (commonly found in tumor cells) can use different modes of microtubule association to regulate distinct aspects of microtubule function in mitosis.

Use of human colorectal tumor cell lines with mutated or wild type APC allowed for the role of APC in mitosis to be more directly evaluated. The majority of mutations associated with familial or sporadic colorectal cancer are nonsense or frameshift mutations that introduce a stop codon and truncate APC, eliminating the central repeat that interacts with β-catenin and the carboxy terminus that interacts with microtubules. When tumor cell lines with chromosome instability (CIN+) were compared to tumor cell lines with stable chromosome inheritance (CIN−; see Table 1), CIN+ tumor cells were found to have a poorly organized mitotic apparatus, fewer astral microtubules that contact the cell cortex and less stable kinetochore-microtubule attachments. These observations suggest that changes at the plus-ends of microtubules might underlie the disorganized mitotic apparatus (Fig. 1).[13] Although CIN+ cells all exhibited a similar array of

Table 1. ***A summary of the liteature is shown to summarize colon cancer cell lines that were assayed for chromosomal instability (CIN+or–), aneuploidy and/or microsatellite instability (MIN+or–). CIN status is only + if a rate of chromosome loss has been measured. Similarly, mitotic spindle defect status is only designated when spindle staining has been reported. Aneuploidy is inferred from modal chromosome numbers significantly greater than normal. References are: 1-[13], 2-[75], 3-[76], 4-[15] and 5-[77]. *The first amino-acid codon mutated is listed; the resulting stop codon is further downstream. **Mitotic spindle phenotype is based on the reduction of kinetochore-microtubule attachments, astral microtubules and the appearance of misaligned chromosomes.***

Cell Line	CIN	MIN	APCMut #1*	APCMut #2	Modal Chromo #	Type of Mutation (Allele 1, Allele 2) (APC Mutation Unless Noted)	Mitotic Spindle Defect**	Reference
Caco/ Caco2	CIN+	MIN–	1367	LOH	96	Stop codon	+	1,2
DLD1	CIN–	MIN+	Mutation undefined	LOH	46	N/D	N/D	3,4,5
GP2d	N/D	MIN+	1444	LOH	46	Stop codon	N/D	3
Ht29	CIN+	MIN-	853	1555	90	Stop codon; stop codon	+	1,2,4
Lovo	CIN+	MIN+	1114	1430	49,69	Stop codon;stop codon	+	1,2,3,4,5
LS411	N/D	N/D	789	1556	73	Stop codon; stop codon	N/D	2
SW1417	N/D	MIN–	1450	LOH	70	Stop codon	N/D	2
SW403	N/D	MIN–	1197	1278	64	Stop codon; stop codon	N/D	2
SW480	CIN+	MIN–	1338	LOH	58,90,111	Stop codon	+	1,2,3,4,5
SW620	N/D	MIN–	1338	LOH	48	Stop codon	N/D	3
SW837	CIN+	N/D	1450	LOH	38	Stop codon	N/D	3,4
VACO4A	N/D	MIN–	1354	LOH	62	Stop codon	N/D	3
VACO5	N/D	MIN+	1554	1499	46	Stop codon; stop codon	N/D	3
SW48	N/D	MIN+	WT	WT	46	Beta-catenin	N/D	4
HCT116	CIN–	MIN+	WT	WT	24,65	Beta-catenin	–	1,2,3,4,5
LS174T	N/D	MIN+	WT	WT	47	Beta-catenin	N/D	3
RKO	CIN–	MIN+	WT	WT	46	Other	–	1,4
HCA7	N/D	MIN unclear	WT	WT	43	Other	N/D	3
LIM1863	N/D	MIN–	WT	WT	80	Other	N/D	3

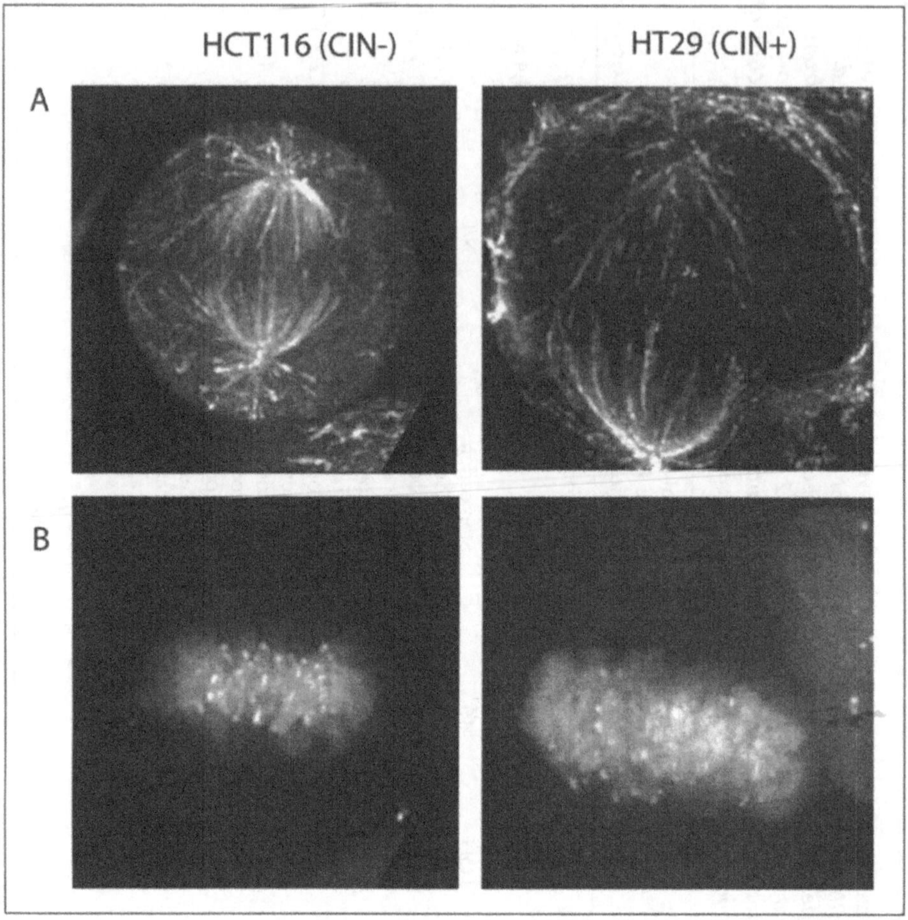

Figure 1. Mitotic spindles in human colorectal cell lines characterized to be CIN– (HCT116) or CIN+ (HT29). Cells were fixed and stained with antibodies against (A) tubulin to show the mitotic spindle and (B) with DAPI and anticentromere antibodies to reveal chromosomes and kinetochores, respectively. Note that the metaphase chromosomes in HT29 cells form a less compact alignment than in HCT116 cells, a defect caused by inefficient kinetochore-microtubule attachment.[13]

mitotic defects, there is a range of severity in different tumor cell lines which suggests that there may well be additional modifiers of mitotic function.[11] Such modifiers, in principle, could allow tumor cells with defective mitoses to overcome otherwise crippling defects in division. Importantly, the mitotic phenotype observed in CIN+ tumor cell lines correlates with the genotype of *APC*; all CIN+ colon tumor cells have mutations in *APC*. Most, but not all, tumor cells with *APC* mutations exhibit CIN as determined directly by nondisjunction events or as inferred by large changes in ploidy (see Table 1). Interestingly, the exceptions (VACO-5, Gp2d and DLD1) exhibit microsatellite instability (MIN), a phenotype that is likely caused by defective DNA mismatch repair. It is tempting to speculate that cells with MIN require modifiers to suppress the mitotic defects associated with *APC* mutations in order to be viable. It is also possible that nondisjunction events occur in these cell lines but the resulting aneuploids are not viable due to the high mutation rate (i.e., loss of heterozygosity). A more careful examination of the mitotic apparatus in these cells will be required to distinguish between these possibilities. These examples not withstanding, there is a strong correlation between mutations in *APC* and the CIN phenotype.

Using fusions of CIN+ and CIN– tumor cells, Vogelstein and colleagues showed that CIN acts dominantly.[15] The dominant quality of CIN allowed for a genetic test of the prediction that *APC* mutations are sufficient to give rise to mitotic defects and to CIN. Conditional expression of a clinically relevant truncation mutant of APC ($APC;^{1\text{-}1450}$ codes for amino acids 1-1450) in human 293 cells (CIN–), which contain wild type APC, resulted in mitotic defects identical to those found in colorectal tumor cell lines; astral microtubules are decreased, kinetochore microtubules are less stable due to inefficient kinetochore attachment and chromosomes are misaligned in metaphase.[13] The dominant activity of $APC^{1\text{-}1450}$ depends on its ability to associate with full length APC via an amino terminal oligomerization domain. Biochemical experiments demonstrated that the dominant negative APC-complex "poisons" the full length APC, preventing it from interacting with EB1. The localization of $APC^{1\text{-}1450}$ at microtubule plus-ends, along the length of spindle microtubules and at the centrosomes, places the dominant negative APC-complex in a position to compromise mitotic spindle function.[13,16] In addition to the amino-terminal oligomerization domain, sequences between amino acids 759 and 1309 contribute to the dominant negative activity of *APC* mutants.[17] Significantly, even very low levels of $APC^{1\text{-}1450}$ perturb spindle function, implying that precancerous cells with one mutant and one wild type copy of APC will have similar mitotic defects. A similar dominant activity was observed when amino terminal fragments of APC were reintroduced into CIN–, HCT116 colorectal tumor cells.[16] Together, these observations provide compelling evidence that APC has an important mitotic function and that it is linked to its association with the microtubule plus-end binding protein, EB1. Obviously, the dominant activity of APC mutants found in cancer cells has important implications for understanding chromosome instability and the genetics of *APC* mutations (discussed below).

APC Regulates Microtubule Dynamics in Mitosis

Dominant alleles of *APC* or siRNA directed reduction of APC protein inhibit spindle microtubule dynamics, compromising the formation of microtubules important for chromosome segregation and for anchoring the spindle to the cell cortex[17] (see cartoon in Fig. 3). To understand the role of APC in mitosis, attention has focused on the microtubule regulatory properties of APC. How does APC regulate microtubule dynamics and is this function distinct from its other roles in the cell? While complete answers to these questions have yet to be found, initial progress has been made by focusing on the carboxy-terminal domains of APC that mediate its interaction with microtubules. Although APC is often described as a microtubule plus-end associated protein (+TIP), its localization along the lengths of some microtubules as well as at plus-ends makes it distinct from other +TIPs.[18] In addition, APC is found only on a subset of cellular microtubules while other +TIPs such as EB1 can be seen on most growing microtubule plus-ends.[19] In migrating cells, APC appears in clusters associated only with the leading edge of the cell.[4] In mitotic cells, as microtubule dynamics increase to facilitate their "search and capture" of mitotic targets such as kinetochores,[20-22] APC is found associated more extensively with a large number of microtubule plus-ends. The more obvious association of APC with spindle microtubules may reflect the increased importance of the microtubule regulatory function of APC in mitosis.

APC has three described modes of microtubule interaction: (i) via the Armadillo repeats (amino acids, 446-880),[23] APC interacts with KAP3, the kinesin light chain associated protein and apparently uses this interaction to move along the length of microtubules; (ii) via its basic domain APC interacts directly with microtubules (amino acids, 2219-2580);[24] and (iii) via a second domain in the carboxy terminus, APC interacts with EB1 (amino acids 2781-2831).[25,26] Localization studies of APC in migrating and mitotic cells are consistent with there being multiple populations of microtubule-associated APC.

The meaning of the diverse localizations of APC in mitosis remains to be determined, however, there are several important implications to be gleaned from our current picture. First, the localization of APC along the lattice of spindle microtubules may indicate that there are discrete pools of APC that are targeted to specific mitotic structures, such as the kinetochore or the cell cortex. Understanding the nature of these complexes and their motility in more detail will be important for determining their significance. Secondly, the presence of APC at the plus-ends of microtubules points to the important relationship between APC and EB1, one that has led to the proposal that this complex regulates microtubule plus-end dynamics. The infrequent interaction between APC and EB1-associated microtubules in interphase cells highlights the complicated relationship between APC and EB1[27] and raises the possibility that APC-EB1 complexes act differently in interphase and mitotic cells. Finally, it is possible that the direct interaction between APC and microtubules leads to changes in the dynamic properties of microtubules. Whatever the relevant biochemical interactions between APC and microtubules, it is likely that the relationship is not simple. For example, a truncated APC mutant (i.e., APC^{1-1450}) forms a hetero-oligomer with the full length APC protein that still associates with microtubule plus-ends, but inhibits the ability of full length APC to interact with EB1. Thus, APC is capable of different modes of interactions with microtubules and its association with EB1 is neither necessary nor sufficient for its localization.[17,28] Ultimately, it will be important to identify binding partners and biochemical activities of the different populations of microtubule-associated APC in order to better understand their functions in both cell migration and mitosis.

Despite the complicated picture of microtubule-associated APC, mutants that disrupt the interaction between APC and EB1 clearly inhibit microtubule dynamics in mitotic cells. As mentioned, APC^{1-1450} acts as a dominant negative by oligomerizing with full length APC and preventing full length APC from interacting with EB1.[17] This biochemical defect correlates with the disorganized mitotic spindles observed in cells expressing APC^{1-1450} and in *APC* mutant tumor cells. A similar phenotype is observed when APC is inhibited by siRNA, thus confirming the dominant negative activity of the APC truncations. Measurements of EB1-comet behavior show that expression of APC^{1-1450} reduces microtubule dynamics by increasing the pause frequency of spindle microtubules. Deletion of the amino-terminal oligomerization domain (i.e., $APC^{58-1450}$) eliminates its dominant negative activity, restores APC-EB1 interaction and restores robust mitotic spindles. Interestingly, the *APC* allele lacking the oligomerization domain acts as a gain-of-function allele, decreasing EB1-comet pause duration relative to controls.[17] That alleles of *APC* demonstrate either loss or gain-of-function phenotypes that correlate with EB1 interaction strongly supports the ability of APC to, directly or indirectly, regulate EB1 and thus microtubule plus-end dynamics.

How then does APC regulate microtubule dynamics? The answer may lie in the ability of APC to simultaneously interact with multiple protein complexes. In this scenario, the association of APC with microtubules could target associated proteins to microtubule plus-ends. For example, the APC-associated proteins, β-catenin, GSK3β and Axin colocalize at microtubule plus-end clusters in migrating epithelial cells,[29] neurons [30,31] and are found together on mitotic spindles (Green and Kaplan; unpublished observations). Thus, it is possible that APC regulates microtubule dynamics through the canonical β-catenin "destruction" complex (a complex that helps target β-catenin for ubiquitin-mediated degradation), perhaps by controlling the activity (i.e., phosphorylation state or protein levels) of other +TIPs. Alternatively, there may be additional APC-binding partners important for regulating microtubule dynamics as suggested by biochemical fractionation experiments which have identified destruction complex free pools of APC associated with membrane clusters.[32-34]

Interestingly, the interaction of other proteins with EB1 suggests that APC is not unique in its ability to bridge interactions between microtubules and complexes important for cell polarity. In this regard, it is instructive to consider the case of melanophilin, a protein that interacts with EB1 and links myosin on melanosomes to microtubule plus-ends. Recent work shows that melanophilin contains an EB1 binding domain with homology to the one found in APC. Genetic studies indicate that the EB1-binding domain in melanophilin links it to microtubule plus-ends in order to deliver melanosomes to the periphery of melanocytes.[35] Such a polarized delivery role is distinct from the more regulatory role proposed for APC at microtubule plus-ends. This difference may not be surprising given the lack of sequence homology outside the EB1-binding domain. Nonetheless, it may be useful to consider the possibility that these cases are more similar than they first appear. For instance, the interaction between APC and EB1 may be necessary to deliver APC to its site of action. On the other hand, the role of melanophilin in microtubule regulation has not been examined; it may be that it feeds back and regulates microtubule plus-ends as it is being transported to the cell periphery. Such a complex relationship would be consistent with functional studies linking APC and complexes important for actin regulation to microtubule plus-ends.[17,36-40]

APC Function in the Spindle Checkpoint

The role of APC in the spindle checkpoint is an important issue not only because it has the potential to inform our understanding of checkpoint function, but also because *APC* mutants that are defective in sensing misaligned chromosomes fulfill a basic prediction of models of genomic instability (i.e., loss of checkpoint function; see Fig. 3 model).[41-43] Spindle checkpoint proteins are found associated with kinetochores that have not been properly attached to the mitotic spindle. Here, they rapidly exchange with a pool of cytosolic checkpoint complexes important for transmitting a "stop-anaphase" signal. Following proper kinetochore-microtubule attachment, spindle checkpoint proteins dissociate from the kinetochore and anaphase proceeds.[44] Although not a prototypical checkpoint protein, there is significant evidence that APC is important for detecting misaligned chromosomes. However, as in its involvement in microtubule dynamics, the precise function of APC in the checkpoint defies simple explanation.

Accumulated evidence suggests that APC functions independently from traditional checkpoint complexes to detect misaligned chromosomes. For example, APC does not accumulate at unattached kinetochores; in fact, it requires microtubules for its kinetochore localization. Moreover, a number of groups have reported that cells defective for APC have "normal" spindle checkpoint function as defined by the ability of cells to arrest in response to microtubule depolymerization.[16,17,45] On the surface, these results indicate that APC is not required for the spindle checkpoint complex to arrest cells. However, expression of the dominant negative APC^{1-1450} in cells induces a dramatic increase in misaligned chromosomes that does not result in mitotic arrest,[17] a finding that suggests these misaligned chromosomes are "invisible" to the spindle checkpoint. This result is particularly suggestive as it is this heterozygote genotype that often precedes colorectal cancer and raises the possibility that APC can simultaneously inhibit chromosome segregation and the quality control mechanisms that prevent chromosome instability. Finally and perhaps most significantly, live-imaging of cells expressing dominant negative mutants of APC (e.g., APC^{1-1450} and N-APC (amino acids 1-750)) or of cells inhibited for APC by siRNA demonstrate a premature exit from mitosis in the presence of lagging chromosomes.[46] The unavoidable conclusion from these studies is that APC assists the spindle checkpoint in detecting misaligned chromosomes.

There is precedent for how misaligned chromosomes can become "invisible" to the spindle checkpoint. Chromosomes that have attached aberrantly to the mitotic spindle so that one of the two attached sisters are connected to microtubules from both poles, known as merotelic attachment, results in tension that inactivates the checkpoint even though chromosomes cannot be segregated to opposite poles.[47] However, the fact that sister kinetochores of misaligned chromosomes in APC defective cells exhibit reduced tension rules out a simple merotelic attachment explanation for why misaligned chromosomes are "invisible".[17,45] Another intriguing possibility is that APC is required to amplify the "stop-anaphase" signal in the context of transient kinetochore-microtubule attachments. Such signal amplification might be important in the presence of nascent kinetochore-microtubule connections which probably give rise to the transient establishment of tension, a state consistent with the pre-anaphase "breathing" observed at centromeres in both human and yeast cells.[48,49] Importantly, such an amplifier would not be required when cells are treated with microtubule poisons that eliminate all microtubules and therefore explains why the checkpoint appears functional when APC defective cells are treated with nocodazole. How does APC amplify a microtubule-dependent checkpoint signal? Interestingly, inhibiting EB1 gives rise to similar "invisible" misaligned chromosomes and therefore argues that the role of APC in amplifying the spindle checkpoint signal is related to its ability to regulate microtubule dynamics.[45] Since inhibiting either EB1 or APC results in less dynamic microtubules that spend more time in a paused state (i.e., neither growing nor shrinking), it may be that maintaining dynamic microtubules at the nascent kinetochore-microtubule interface is important to amplify checkpoint signals prior to full kinetochore-microtubule attachment and maximal force generation. Such a role is consistent with the more general function of APC and EB1 in regulating microtubule plus-end behavior. To address such a model, it will be important to determine how changes in microtubule plus-end behavior impact the activity of the spindle checkpoint complexes.

APC and Microtubule Dynamics in Cytokinesis

Classic experiments by Ray Rappaport established the role of the mitotic spindle in the specification of the cytokinetic furrow.[50] While the specifics vary widely among species, it is clear that a combination of dynamic cortical microtubules, stable centrosomes and midzone microtubules direct actin polymerization and thus the location of the cytokinetic furrow. While the importance of microtubules and actin in cytokinesis are indisputable, the molecular details connecting these two cytoskeleton systems remain less clear. Current models propose that the Rho family of G proteins (i.e., Cdc42, Rac, or Rho) that locally control actin polymerization at the cortex are affected by the action of spindle microtubule dynamics per se or by the ability of dynamic spindle microtubules to "deliver" important regulatory factors (e.g., GAPs and GEFs). The critical role of spindle microtubules in cytokinesis raises the possibility that APC is also important for cytokinesis, either through its influence on anaphase spindle organization, dynamics or microtubule associated activities that influence actin behavior at the cell cortex.

Observations of anaphase spindles in cells defective for APC function provide a strong piece of circumstantial evidence that APC is required for proper cytokinesis. Anaphase cells expressing the dominant negative APC^{1-1450} exhibit a dramatic loss of cortical spindle microtubules, fewer microtubule plus-ends contacting the cortex and an increase in off-center mitotic spindles (Fig. 2).[13] Furthermore, live-cell imaging of cells expressing APC^{1-1450} or of cells inhibited for APC function by siRNA show spindles that undergo dramatic rotational oscillations.[45,51] If we accept the idea that the mitotic spindle acts as a signaling platform whose position can influence cytokinetic furrow induction, [52] then it is compelling to ask what might happen to cytokinesis in cells that fail to anchor their mitotic spindle. An important clue comes from studies in *Xenopus laevis*, where mechanically sliding the mitotic spindle along the cortex of the cell expands the zone of active RhoA and actin polymerization.[53] Consistent with this result, we have observed that epithelial cells with defective spindle anchoring caused by mutant APC have a similarly expanded zone of active RhoA (Caldwell and Kaplan; unpublished observations). It should be noted, however, that APC function is not required for cytokinesis in Drosophila.[54] Therefore, at least in mammalian systems, the ability of APC to regulate anaphase spindle function appears to be important for mitotic cells to properly organize their cortex for cytokinesis.

If these types of changes in anaphase spindle behavior severely compromise cytokinesis then it follows that cells with reduced APC function will become polyploid (i.e., tetraploid) due to cytokinetic failure. Evidence to support this prediction comes from the observation that cells expressing APC^{1-1450} have a significant number of anaphases where there is no observable cytokinetic furrow; live-cell imaging reveals that such cells exit mitosis with normal timing and thus give rise to bi- or multi-nucleated cells.[51] In addition, depletion of APC in cultured cells rapidly leads to the accumulation of tetra- and polyploid cells.[46] Importantly, not every cell expressing APC^{1-1450} fails to anchor its spindle or fails to complete cytokinesis, implying that either APC dominant mutants are not null or that there are multiple pathways that contribute to spindle anchoring, or both. Significantly, this observation also implies that tumor cells that have lost APC function can still undergo successful cell division and therefore are able to contribute to the expansion of the tumor. Similar polyploid cells have been recently reported for mice containing a targeted knockout of APC function[46] and agree with our data that dominant *APC* alleles contribute to polyploidy in intestinal epithelial cells.[51] Together, these observations imply that cytokinetic failures in cells

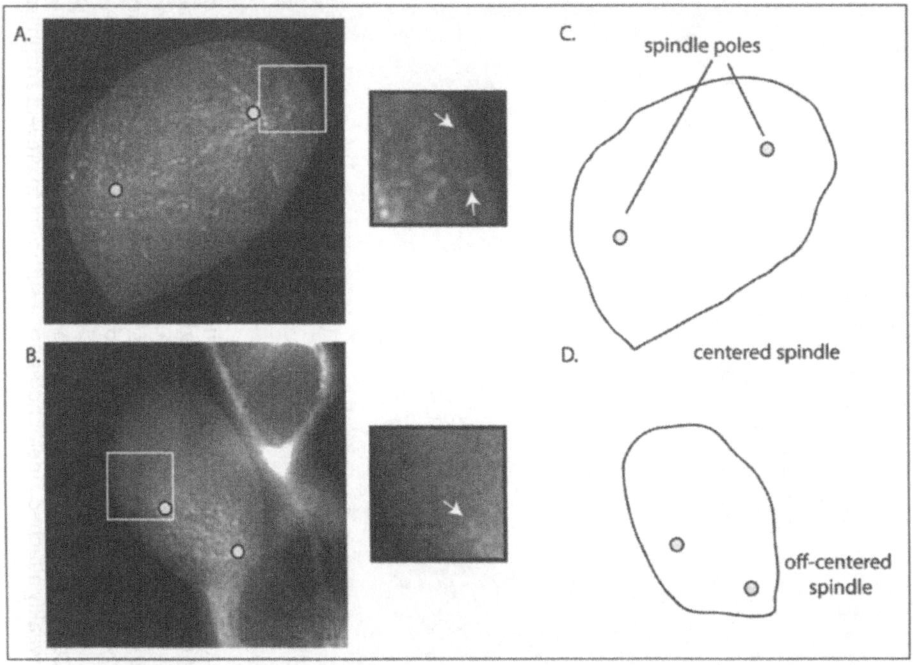

Figure 2. Expression of the clinically relevant APC$^{1-1450}$ in cells with wild type APC prevents normal organization of anaphase spindles. Control human 293 cells (A) or cells conditionally expressing APC$^{1-1450}$ (B) were transfected with EB1-GFP to visualize the mitotic spindle and microtubule plus-ends (see arrows). The region inside the box was enlarged 1.5-fold to show microtubule plus-ends near the cortex. Yellow filled in circles indicate the approximate position of the spindle poles. (C) and (D) are cartoons derived from the cells in (A) and (B) to show the relative position of the spindle poles with respect to the cell boundary. A color version of this figure is available online at www.landesbioscience.com/curie.

that lose APC function are not simply an in vitro artifact of cell culture and that mitotic defects occur in tissues where *APC* mutations give rise to cancer.

How might APC contribute to anaphase spindle signals required for cytokinetic furrow induction? There are two general possibilities: (i) APC influence over signals generated from anaphase spindles may be indirect, simply reflecting the need for normal microtubule dynamics necessary to bring cortical microtubules to the cortex or, (ii) APC may directly form or regulate signals that control actin dynamics at the cortex (see model, Fig. 3). A variety of experiments argue that microtubule dynamics per se are not required for successful cytokinesis[55,56] and thus point to a role for microtubule plus-end binding complexes in regulating actin at the cortex. More direct evidence for such a relationship comes from the observed connection between APC, EB1 and the RhoA effector, mDia (a formin that stimulates actin polymerization) in migrating fibroblasts, a relationship that suggests the possibility of directly linking microtubule-associated APC complexes to actin during cytokinesis.[7] While this complex has been proposed to stabilize microtubules in migrating cells, evidence from sea urchin embryos suggest that complexes on spindle microtubule plus-ends, including APC, EB1 and p150glued, are necessary for proper cytokinetic timing.[57] In addition to the cytoskeleton-related APC complexes, it is also important to consider how the well-characterized β-catenin "destruction" complex may influence actin regulation. For example, APC2, GSK3β, Armadillo and α-catenin have been shown to be important for spindle anchoring in Drosophila syncytial mitoses.[58] In addition, APC forms a complex with Axin-GSK3β and

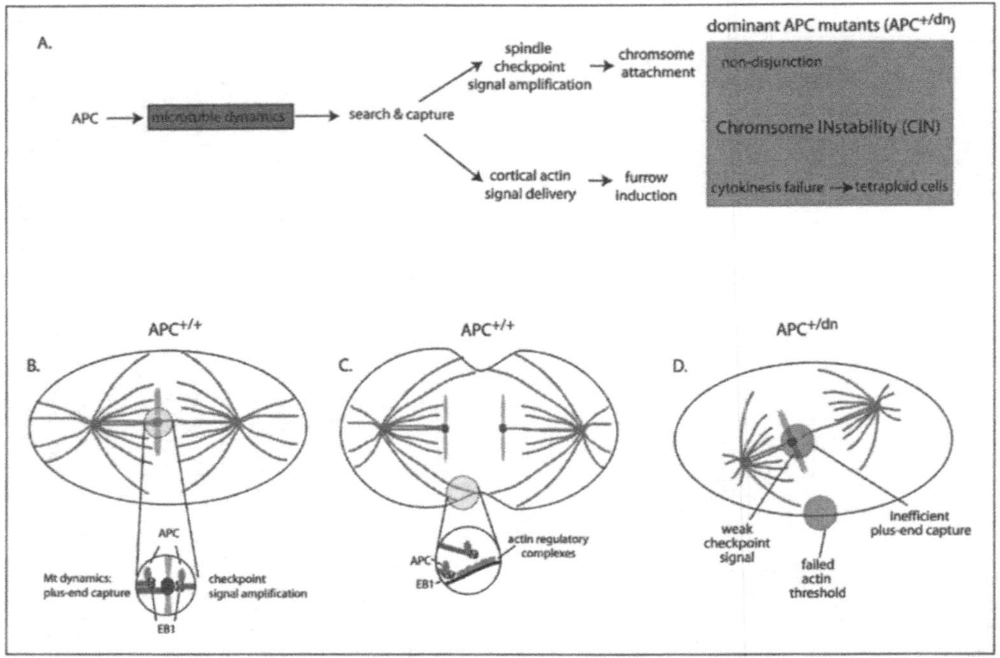

Figure 3. APC is connected to multiple mitotic processes through its ability to regulate microtubule dynamics. A) The flow chart connects the role of APC in regulating microtubule dynamics to the ability of cells to properly attach chromosomes to the mitotic spindle and to induce the cytokinetic furrow. Phenotypes associated with the dominant negative (dn) alleles of *APC* are highlighted in the grayed box. B) Alignment of chromosomes during metaphase requires efficient kinetochore-microtubule attachments. APC, through its interaction with EB1, influences microtubule dynamics necessary for efficient "search and capture" and we propose serves a secondary role in amplifying checkpoint signals initiated at newly attached kinetochores. C) Cortical microtubules depend on their proximity to the anaphase cortex to regulate actin polymerization associated with furrow induction. This regulation appears to require multiple microtubule plus-end proteins, including APC and EB1. D) In cells expressing a single dominant negative mutant of *APC*, kinetochore-microtubule attachments are reduced, checkpoint signals from weakly attached kinetochores are insufficient to prevent anaphase and cortical microtubules are unable to properly regulate actin at the cell cortex.

MCAF1, a microtubule-actin cross linking factor.[59] It is therefore possible that APC-microtubule associated complexes integrate with those complexes known to regulate β-catenin both during changes in cell polarity and during anaphase to complete cytokinesis. It will be necessary to define the composition of APC complexes in mitotic cells to identify activities required for the initiation of the cytokinetic furrow (as summarized in Fig. 3B).

The Dominant Activity of APC in Mitosis-The Tip of the Iceberg

While studies that have examined the cell biology of APC have informed our view of its role in cell polarity and mitosis, the association between APC and colorectal cancer in humans adds both significance as well as an important dimension to the discussion. Specifically, it is significant that APC mutants found in human cancer act as dominant negatives to inhibit microtubule dynamics. Inheritance or the sporadic occurrence of such alleles can now be connected to specific cellular phenotypes and therefore informs our view of the long-described dominant predisposition to cancer in familial adenomatous polyposis (FAP) patients inheriting one mutant allele of *APC* (see Fig. 3A). Indeed, the majority of FAP patients have a mutation in *APC* resulting in the expression of a truncated protein. FAP patients are assured of developing hundreds to thousands of colonic polyps with adenomas often appearing in preteen and young adult years.[60] The question that naturally arises is: does the defect in spindle microtubules contribute to the predisposition of colorectal cancer? By considering the answer to this question, we can begin to appreciate how the observed mitotic defects represent the tip of a phenotypic iceberg-a subset, or all of which, may help to "evolve" tumor cells in the gut.

Ultimately, to address this question we must understand the full range of cellular phenotypes associated with the reduction of microtubule dynamics caused by dominant *APC* mutants. While the global role of microtubules in organizing epithelial cells raises numerous potential consequences for cells, we will focus on two that could in principle drive cancer cell evolution: (i) the dominant effect of APC on microtubule dynamics may impact the normal organization of polarized epithelial cells with respect to cell migration as well as mitotic spindle orientation and (ii), changes in mitotic fidelity may contribute to genomic (i.e., chromosome) instability. We will briefly address each of these possibilities and the ramifications of such changes for cancer progression. The dynamic exchange of APC between complexes that degrade β-catenin and complexes that bind to microtubules argues that there is a direct relationship between microtubule dynamics and the stability of adherence junctions.[34,61,62] In addition, a number of studies have linked APC, EB1 and microtubules to events during cell migration such as the stabilization of microtubules at focal contacts.[7] While the molecular connection between microtubules and actin remains unclear, visualization of these structures in living cells has highlighted the relationship between growing microtubules and actin polymerization.[27,63,64] It is therefore conceivable that, as in cytokinesis, changes in microtubule dynamics in interphase cells could contribute to the well-documented changes in epithelial cell adhesion and migration observed in tumors with *APC* mutations.[65] In many epithelial tissues, the case has been made for the importance of spindle orientation in the asymmetric cell divisions that allow stem cell maintenance and cell differentiation.[66,67] Changes in spindle orientation in the gut could therefore have significant impact on the properties of intestinal stem cells. Importantly, all of these changes are postulated to occur prior to LOH at the *APC* locus and may expose intestinal epithelial cells to new selective pressures in the intestinal crypt by causing changes in cell position, stem cell maintenance and loss of proper signals that normally restrain cell division.

As reviewed above, the evidence is compelling that the dominant activity of APC mutants contributes to mitotic defects in chromosome alignment and cytokinesis both in vitro and in mice. While many tumors exhibit aneuploidy, it remains controversial whether these changes cause cancer (onset or progression) or, instead, are an effect of accumulated lesions present in tumor cells. The dominant nature of APC truncating mutants argues that mitotic errors can occur prior to other changes associated with colorectal cancer (i.e., LOH of APC and stabilization of β-catenin). Thus, the early presentation of the phenotype might persuade some that mitotic defects are important at the earliest stages of cancer. But what evidence is there that changes in mitotic fidelity can cause cancer? The long-cited correlation between aneuploidy and malignancy minimally suggests that changes in chromosome number may be a late event that enhances characteristics associated with malignancy (e.g., metastasis or angiogenesis).[68] However, polyploidy may reflect a more global genomic stability problem and has been proposed to be a driving force in cancer.[69] In support of this contention, studies in yeast demonstrate that polyloid cells are more susceptible to chromosome loss and DNA repair defects.[70,71] Parallel experiments in mammalian cells showed that artificially creating polyploid cells promote mitotic errors and tumor formation.[72,73] The occurrence of tetraploid cells in multiple human cancers has led to the model that tetraploid cell formation is a rate limiting event that precedes other forms of genomic instability and ultimately cancer progression.[74] The significant rate of cytokinetic failure in cultured cells and in mice containing a single mutant allele of APC provides the first direct evidence that a clinically relevant lesion can promote tetraploid formation early in cancer, thus placing it as a cause rather than effect of cancer. In the future, it will be necessary to use mouse models to test how the early mitotic defects associated with APC mutants contribute to colorectal cancer progression.

References

1. Siegrist SE, Doe CQ. Microtubule-induced cortical cell polarity. Genes Dev 2007; 21(5):483-496.
2. Basu R, Chang F. Shaping the actin cytoskeleton using microtubule tips. Curr Opin Cell Biol 2007; 19(1):88-94.
3. Etienne-Manneville S, Hall A. Cdc42 regulates GSK-3beta and adenomatous polyposis coli to control cell polarity. Nature 13 2003; 421(6924):753-756.
4. Nathke IS, Adams CL, Polakis P et al. The adenomatous polyposis coli tumor suppressor protein localizes to plasma membrane sites involved in active cell migration. J Cell Biol 1996; 134(1):165-179.
5. Mimori-Kiyosue Y, Shiina N, Tsukita S. The dynamic behavior of the APC-binding protein EB1 on the distal ends of microtubules. Curr Biol 2000; 10(14):865-868.
6. Yamana N, Arakawa Y, Nishino T et al. The Rho-mDia1 pathway regulates cell polarity and focal adhesion turnover in migrating cells through mobilizing Apc and c-Src. Mol Cell Biol 2006; 26(18):6844-6858.
7. Wen Y, Eng CH, Schmoranzer J et al. EB1 and APC bind to mDia to stabilize microtubules downstream of Rho and promote cell migration. Nat Cell Biol 2004; 6(9):820-830.
8. Gundersen GG, Gomes ER, Wen Y. Cortical control of microtubule stability and polarization. Curr Opin Cell Biol 2004; 16(1):106-112.
9. Cleveland DW, Mao Y, Sullivan KF. Centromeres and kinetochores: from epigenetics to mitotic checkpoint signaling. Cell 2003; 112(4):407-421.
10. Kline-Smith SL, Sandall S, Desai A. Kinetochore-spindle microtubule interactions during mitosis. Curr Opin Cell Biol 2005; 17(1):35-46.
11. Fodde R, Kuipers J, Rosenberg C et al. Mutations in the APC tumour suppressor gene cause chromosomal instability. Nat Cell Biol 2001; 3(4):433-438.
12. Kaplan KB, Burds AA, Swedlow JR et al. A role for the Adenomatous Polyposis Coli protein in chromosome segregation. Nat Cell Biol 2001; 3(4):429-432.
13. Green RA, Kaplan KB. Chromosome instability in colorectal tumor cells is associated with defects in microtubule plus-end attachments caused by a dominant mutation in APC. J Cell Biol 2003; 163(5):949-961.
14. Zumbrunn J, Kinoshita K, Hyman AA et al. Binding of the adenomatous polyposis coli protein to microtubules increases microtubule stability and is regulated by GSK3 beta phosphorylation. Curr Biol 2001; 11(1):44-49.
15. Lengauer C, Kinzler KW, Vogelstein B. Genetic instability in colorectal cancers. Nature 1997; 386(6625):623-627.
16. Tighe A, Johnson VL, Taylor SS. Truncating APC mutations have dominant effects on proliferation, spindle checkpoint control, survival and chromosome stability. J Cell Sci 2004; 117(Pt 26):6339-6353.
17. Green RA, Wollman R, Kaplan KB. APC and EB1 Function Together in Mitosis to Regulate Spindle Dynamics and Chromosome Alignment. Mol Biol Cell 2005; 16(10):4609-4622.
18. Carvalho P, Tirnauer JS, Pellman D. Surfing on microtubule ends. Trends Cell Biol 2003; 13(5):229-237.
19. Piehl M, Tulu US, Wadsworth P et al. Centrosome maturation: measurement of microtubule nucleation throughout the cell cycle by using GFP-tagged EB1. Proc Natl Acad Sci USA 2004; 101(6):1584-1588.
20. Holy TE, Leibler S. Dynamic instability of microtubules as an efficient way to search in space. Proc Natl Acad Sci USA 1994; 91(12):5682-5685.

21. Schuyler SC, Pellman D. Search, capture and signal: games microtubules and centrosomes play. J Cell Sci 2001; 114(Pt 2):247-255.
22. Wollman R, Cytrynbaum EN, Jones JT et al. Efficient chromosome capture requires a bias in the 'search-and-capture' process during mitotic-spindle assembly. Curr Biol 2005; 15(9):828-832.
23. Jimbo T, Kawasaki Y, Koyama R et al. Identification of a link between the tumour suppressor APC and the kinesin superfamily. Nat Cell Biol 2002; 4(4):323-327.
24. Deka J, Kuhlmann J, Muller O. A domain within the tumor suppressor protein APC shows very similar biochemical properties as the microtubule-associated protein tau. Eur J Biochem 1998; 253(3):591-597.
25. Bu W, Su LK. Characterization of functional domains of human EB1 family proteins. J Biol Chem 2003; 278(50):49721-49731.
26. Slep KC, Rogers SL, Elliott SL et al. Structural determinants for EB1-mediated recruitment of APC and spectraplakins to the microtubule plus end. J Cell Biol 2005; 168(4):587-598.
27. Kita K, Wittmann T, Nathke IS et al. Adenomatous polyposis coli on microtubule plus ends in cell extensions can promote microtubule net growth with or without EB1. Mol Biol Cell 2006; 17(5):2331-2345.
28. Barth AI, Siemers KA, Nelson WJ. Dissecting interactions between EB1, microtubules and APC in cortical clusters at the plasma membrane. J Cell Sci 2002; 115(Pt 8):1583-1590.
29. Sharma M, Leung L, Brocardo M et al. Membrane localization of adenomatous polyposis coli protein at cellular protrusions: targeting sequences and regulation by beta-catenin. J Biol Chem 2006; 281(25):17140-17149.
30. Ciani L, Krylova O, Smalley MJ et al. A divergent canonical WNT-signaling pathway regulates microtubule dynamics: Dishevelled signals locally to stabilize microtubules. J Cell Biol 2004; 164(2):243-253.
31. Votin V, Nelson WJ, Barth AI. Neurite outgrowth involves adenomatous polyposis coli protein and beta-catenin. J Cell Sci 2005; 118(Pt 24):5699-5708.
32. Mahadevaiyer S, Xu C, Gumbiner BM. Characterization of a 60S complex of the adenomatous polyposis coli tumor suppressor protein. Biochim Biophys Acta 2007; 1773(2):120-130.
33. Reinacher-Schick A, Gumbiner BM. Apical membrane localization of the adenomatous polyposis coli tumor suppressor protein and subcellular distribution of the beta-catenin destruction complex in polarized epithelial cells. J Cell Biol 2001; 152(3):491-502.
34. Penman GA, Leung L, Nathke IS. The adenomatous polyposis coli protein (APC) exists in two distinct soluble complexes with different functions. J Cell Sci 2005; 118(Pt 20):4741-4750.
35. Wu XS, Tsan GL, Hammer JA 3rd. Melanophilin and myosin Va track the microtubule plus end on EB1. J Cell Biol 2005; 171(2):201-207.
36. Kawasaki Y, Senda T, Ishidate T et al. Asef, a link between the tumor suppressor APC and G-protein signaling. Science 2000; 289(5482):1194-1197.
37. Rosin-Arbesfeld R, Ihrke G, Bienz M. Actin-dependent membrane association of the APC tumour suppressor in polarized mammalian epithelial cells. EMBO J 2001; 20(21):5929-5939.
38. Kawasaki Y, Sato R, Akiyama T. Mutated APC and Asef are involved in the migration of colorectal tumour cells. Nat Cell Biol 2003; 5(3):211-215.
39. Ligon LA, Shelly SS, Tokito M et al. The microtubule plus-end proteins EB1 and dynactin have differential effects on microtubule polymerization. Mol Biol Cell 2003; 14(4):1405-1417.
40. Watanabe T, Wang S, Noritake J et al. Interaction with IQGAP1 links APC to Rac1, Cdc42 and actin filaments during cell polarization and migration. Dev Cell 2004; 7(6):871-883.
41. Wassmann K, Benezra R. Mitotic checkpoints: from yeast to cancer. Curr Opin Genet Dev 2001; 11(1):83-90.
42. Draviam VM, Xie S, Sorger PK. Chromosome segregation and genomic stability. Curr Opin Genet Dev 2004; 14(2):120-125.
43. Jallepalli PV, Lengauer C. Chromosome segregation and cancer: cutting through the mystery. Nat Rev Cancer 2001; 1(2):109-117.
44. Taylor SS, Scott MI, Holland AJ. The spindle checkpoint: a quality control mechanism which ensures accurate chromosome segregation. Chromosome Res 2004; 12(6):599-616.
45. Draviam VM, Shapiro I, Aldridge B et al. Misorientation and reduced stretching of aligned sister kinetochores promote chromosome missegregation in EB1- or APC-depleted cells. EMBO J 2006; 25(12):2814-2827.
46. Dikovskaya D, Schiffmann D, Newton IP et al. Loss of APC induces polyploidy as a result of a combination of defects in mitosis and apoptosis. J Cell Biol 2007; 176(2):183-195.
47. Salmon ED, Cimini D, Cameron LA et al. Merotelic kinetochores in mammalian tissue cells. Philos Trans R Soc Lond B Biol Sci 2005; 360(1455):553-568.
48. Shelby RD, Hahn KM, Sullivan KF. Dynamic elastic behavior of alpha-satellite DNA domains visualized in situ in living human cells. J Cell Biol 1996; 135(3):545-557.
49. He X, Rines DR, Espelin CW et al. Molecular analysis of kinetochore-microtubule attachment in budding yeast. Cell 2001; 106(2):195-206.

50. Rappaport R. Establishment of the mechanism of cytokinesis in animal cells. Int Rev Cytol 1986; 105:245-281.
51. Caldwell CM, Green RA, Kaplan KB. APC mutations lead to cytokinetic failures in vitro and tetraploid genotypes in Min mice. J Cell Biol 2007; 178(7):1109-1120.
52. Rappaport R. Cleavage furrow establishment by the moving mitotic apparatus. Dev Growth Differ 1997; 39(2):221-226.
53. Bement WM, Benink HA, von Dassow G. A microtubule-dependent zone of active RhoA during cleavage plane specification. J Cell Biol 2005; 170(1):91-101.
54. McCartney BM, Price MH, Webb RL et al. Testing hypotheses for the functions of APC family proteins using null and truncation alleles in Drosophila. Development 2006; 133(12):2407-2418.
55. Strickland LI, Donnelly EJ, Burgess DR. Induction of cytokinesis is independent of precisely regulated microtubule dynamics. Mol Biol Cell 2005; 16(10):4485-4494.
56. Shannon KB, Canman JC, Ben Moree C et al. Taxol-stabilized microtubules can position the cytokinetic furrow in mammalian cells. Mol Biol Cell 2005; 16(9):4423-4436.
57. Strickland LI, Wen Y, Gundersen GG et al. Interaction between EB1 and p150glued is required for anaphase astral microtubule elongation and stimulation of cytokinesis. Curr Biol 2005; 15(24):2249-2255.
58. McCartney BM, McEwen DG, Grevengoed E et al. Drosophila APC2 and Armadillo participate in tethering mitotic spindles to cortical actin. Nat Cell Biol 2001; 3(10):933-938.
59. Chen HJ, Lin CM, Lin CS et al. The role of microtubule actin cross-linking factor 1 (MACF1) in the Wnt signaling pathway. Genes Dev 2006; 20(14):1933-1945.
60. Lynch HT, de la Chapelle A. Hereditary colorectal cancer. N Engl J Med 2003; 348(10):919-932.
61. Bienz M. The subcellular destinations of APC proteins. Nat Rev Mol Cell Biol 2002; 3(5):328-338.
62. Faux MC, Ross JL, Meeker C et al. Restoration of full-length adenomatous polyposis coli (APC) protein in a colon cancer cell line enhances cell adhesion. J Cell Sci 2004; 117(Pt 3):427-439.
63. Waterman-Storer CM, Worthylake RA, Liu BP et al. Microtubule growth activates Rac1 to promote lamellipodial protrusion in fibroblasts. Nat Cell Biol 1999; 1(1):45-50.
64. Kroboth K, Newton IP, Kita K, et al. Lack of adenomatous polyposis coli protein correlates with a decrease in cell migration and overall changes in microtubule stability. Mol Biol Cell 2006; 18(3),910-918.
65. Nathke IS. The adenomatous polyposis coli protein: the Achilles heel of the gut epithelium. Annu Rev Cell Dev Biol 2004; 20:337-366.
66. Yamashita YM, Jones DL, Fuller MT. Orientation of asymmetric stem cell division by the APC tumor suppressor and centrosome. Science 2003; 301(5639):1547-1550.
67. Roegiers F, Jan YN. Asymmetric cell division. Curr Opin Cell Biol 2004; 16(2):195-205.
68. Zimonjic D, Brooks MW, Popescu N et al. Derivation of human tumor cells in vitro without widespread genomic instability. Cancer Res 2001; 61(24):8838-8844.
69. Duesberg P, Li R. Multistep carcinogenesis: a chain reaction of aneuploidizations. Cell Cycle 2003; 2(3):202-210.
70. Lin H, de Carvalho P, Kho D et al. Polyploids require Bik1 for kinetochore-microtubule attachment. J Cell Biol 2001; 155(7):1173-1184.
71. Storchova Z, Breneman A, Cande J et al. Genome-wide genetic analysis of polyploidy in yeast. Nature 2006; 443(7111):541-547.
72. Shi Q, King RW. Chromosome nondisjunction yields tetraploid rather than aneuploid cells in human cell lines. Nature 2005; 437(7061):1038-1042.
73. Fujiwara T, Bandi M, Nitta M et al. Cytokinesis failure generating tetraploids promotes tumorigenesis in p53-null cells. Nature 2005; 437(7061):1043-1047.
74. Storchova Z, Pellman D. From polyploidy to aneuploidy, genome instability and cancer. Nat Rev Mol Cell Biol 2004; 5(1):45-54.
75. Rowan AJ, Lamlum H, Ilyas M et al. APC mutations in sporadic colorectal tumors: A mutational "hotspot" and interdependence of the "two hits". Proc Natl Acad Sci USA 2000; 97(7):3352-3357.
76. Abdel-Rahman WM, Katsura K, Rens W et al. Spectral karyotyping suggests additional subsets of colorectal cancers characterized by pattern of chromosome rearrangement. Proc Natl Acad Sci USA 2001; 98(5):2538-2543.
77. Haigis KM, Caya JG, Reichelderfer M et al. Intestinal adenomas can develop with a stable karyotype and stable microsatellites. Proc Natl Acad Sci USA 2002; 99(13):8927-8931.

CHAPTER 6

Role of APC and Its Binding Partners in Regulating Microtubules in Mitosis

Shirin Bahmanyar, W. James Nelson and Angela I.M. Barth*

Abstract

Adenomatous polyposis coli (APC) is a multifunctional protein commonly mutated in colon cancer. APC contains binding sites for multiple proteins with diverse roles in signaling and the structural and functional organization of cells. Recent evidence suggests roles for APC and some of its binding partners in regulating microtubules in mitosis. APC localizes to three key locations in mitosis: kinetochores, the cortex and centrosomes. Here, we discuss possible mechanisms for APC function at these sites and suggest new pathways by which APC mutations promote tumorigenesis.

Introduction

Mutations in *adenomatous polyposis coli* (*APC*) are associated with most colon cancers[1,2] although how these mutations affect the development of cancer is not fully understood. In part, this is because APC is a multifunctional protein involved in a wide variety of cellular processes, as indicated by the number of APC interacting proteins (Fig. 1A). A subset of these interacting proteins, β-catenin and Axin, form a complex that regulates Wnt signaling.[3-7] Other interacting proteins, such as Kinesin-Associated Protein 3 (Kap3), Mitotic Centromere Associated Kinesin (MCAK), mDia, microtubules (MT) and End-Binding 1 (EB1) appear to play a role in microtubule dynamics.[8-12] Studies have shown involvement of APC in cellular functions related to microtubule dynamics such as migration of epithelial cells and neuronal growth cones.[13,14] Here we focus on the role of APC and its binding partners in regulating microtubules in mitosis. APC has been reported to act at three key locations for normal progression through mitosis: the kinetochore, cortex and centrosome.[15-19] Potential roles for APC in mitosis suggest new pathways by which *APC* mutations can contribute to cancer progression.

APC at the Kinetochore: Regulation of Microtubule Dynamics

APC localizes to kinetochores and forms a complex with kinetochore-bound proteins.[15,16] Microtubule-plus end proteins at kinetochores are thought to attach microtubules to chromosomes and/or regulate the local polymerization and depolymerization of microtubules.[20] The opposing stabilizing and destabilizing activities of kinetochore-attached microtubules facilitate chromosome congression at the metaphase plate,[21] which is required for equal and opposite segregation of chromosomes in anaphase.

Roles for APC in either kinetochore-microtubule capture and attachment, or regulation of kinetochore-microtubule plus-end dynamics have been proposed.[22] In support of a role for APC

*Corresponding Author: Angela I.M. Barth—Department of Biological Sciences, Bio-X Program, The James H. Clark Center, 318 Campus Drive (E200), Stanford University, Stanford, California 94305-5430, USA. Email: angelab@stanford.edu

APC Proteins, edited by Inke S. Näthke and Brooke M. McCartney. ©2009 Landes Bioscience and Springer Science+Business Media.

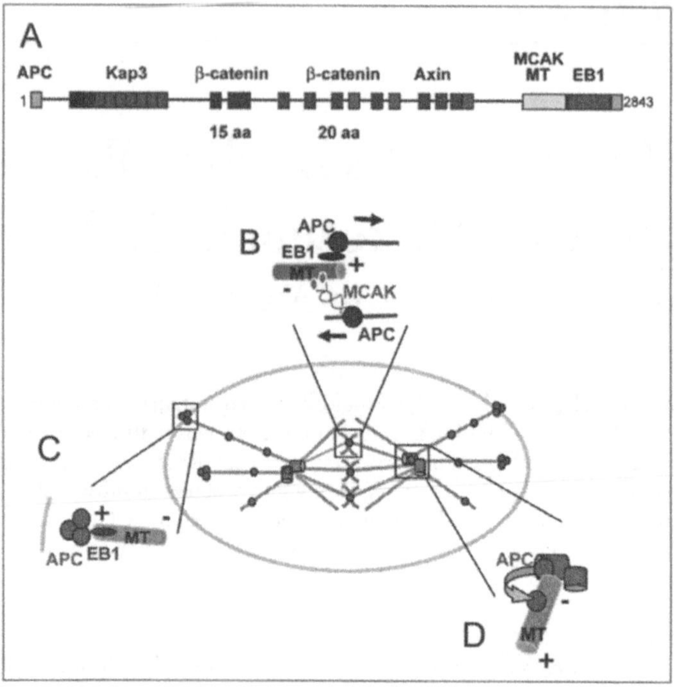

Figure 1. APC regulates microtubules at multiple mitotic locations. a) Domains of vertebrate APC that are involved in microtubule regulation or Wnt signaling are indicated. APC can bind microtubules indirectly through its interaction with Kap3 (Kinesin associated protein 3) via the N-terminal Armadillo repeats (red), or MCAK (Mitotic Centromere Associated Kinesin) and EB1 (End-Binding 1) at the C-terminus. APC can also bind microtubules directly through its microtubule (MT) binding domain. APC has multiple binding sites for β-catenin (green and blue) and Axin (purple), which are located in the central domain. b) APC localizes to kinetochores where it may have key functions in regulating local microtubule plus-end dynamics. APC may coordinate stabilization and destabilization of microtubule plus-ends at kinetochores through its interactions with EB1 and MCAK, respectively. c) In spindle positioning, a proposed role for APC may be to stabilize and attach astral microtubules to the cortex. EB1 may target microtubule plus-ends to cortical APC clusters. Further work is required to determine whether APC localizes in clusters in mitosis. d) APC localizes to the mother centriole where it is proposed to load onto a subset of microtubules with specific functions, such as cortical attachment, spindle orientation and chromosome congression. A color version of this image is available at www.landesbioscience.com/curie.

in microtubule capture and attachment at kinetochores, colon cancer cell lines with mutant *APC* have fewer kinetochores with juxtaposed microtubule ends and an overall decrease in midzone microtubules.[22] However, in contrast to depletion of other microtubule plus-end binding proteins (+Tips) that function in microtubule capture at kinetochores, such as CLIP170, cells depleted of APC have normal kinetochore-attached (cold-stable) microtubules,[23] and decreased levels of Mad2 at kinetochores, which normally accumulates to unattached kinetochores.[23,24] Thus, similar to the role of APC in regulating microtubule plus-end dynamics in migrating cells and mitotic Xenopus extracts,[25-27] an alternative, but not mutually exclusive function for APC at kinetochores could be to regulate local microtubule dynamics. A role for APC in regulating microtubule dynamics at kinetochores is supported by the finding that the distances between kinetochores of sister chromatids is decreased in APC depleted cells.[23,24] This is similar to what happens in response to taxol, which inhibits microtubule dynamics but not attachment at kinetochores.[23,24] Thus, APC depletion results in reduced tension at kinetochores, an indicator of abnormal kinetochore-microtubule dynamics.[28] Interestingly, reduced tension at kinetochores in APC depleted cells correlates with abnormal chromosome congression.[23] Since localization of APC to kinetochores is microtubule-dependent,[16] APC may regulate kinetochore microtubule plus-end dynamics of unattached microtubules and thereby promote microtubule capture at kinetochores. Once the microtubules are attached, APC may regulate local kinetochore microtubule plus-end dynamics to generate tension, which is required for chromosome alignment at the metaphase plate.

APC may stimulate microtubule growth by directly interacting with microtubules through the basic domain or through its interaction with known microtubule stabilizing proteins such as EB1.[9,29] Consistent with the latter idea, out of a panel of +Tips analyzed, EB1 was most related to APC in that its localization to kinetochores was microtubule dependent and depletion of EB1 caused reduced tension but not loss of microtubule attachment at kinetochores.[23] The only noted difference was that EB1 depletion does not engage a spindle checkpoint arrest, whereas APC depletion results in a transient delay in metaphase.[23] Nonetheless, depleting either EB1 or APC causes similar chromosome defects that correlate with and are likely a consequence of, abnormal microtubule dynamics at kinetochores.[23] APC also binds to MCAK, which destabilizes microtubules.[10,30] Thus, APC may coordinate microtubule polymerization and depolymerization at kinetochores allowing for proper chromosome movement and congression at the metaphase plate (Fig. 1B). Future work analyzing how kinetochore proteins that form a complex with APC, such as MCAK and EB1, are affected in APC depleted cells could test this possibility.

APC at the Cortex: Role in Spindle Positioning

A potential role for APC in orienting the mitotic spindle seems to be conserved in several organisms.[31-33] The mechanism of spindle orientation is not well understood, but the importance for microtubule attachment at specialized cortical sites has been established.[31,32,34] In *Saccharomyces cerevisiae*, the functional analogue for APC, Kar9, specifies the cortical site for astral microtubule attachment.[35,36] In *Drosophila*, EAPC/dAPC2 at cell-cell adhesion sites is thought to mediate attachment of astral microtubules at the membrane for proper spindle positioning.[17,18,37,38] Less is known about the role of APC in spindle positioning in mammalian cells although there is evidence to suggest that APC may regulate astral microtubule stability[39] and/or provide a cortical site for astral microtubule attachment (see below).

In the current model in *S. cerevisiae*, Kar9 binds asymmetrically to the daughter-bound spindle pole body (SPB) where it interacts with Bim1 (the yeast homologue of EB1) and is loaded onto the plus-end of a subset of microtubules polymerizing from the SPB (Fig. 2A).[40,41] Kar9-Bim1 bound microtubules are guided along actin cables via Myo2 to the bud tip.[33,42,43] Once attached to the bud tip, microtubule movement can be generated by microtubule plus-end dynamics against the cell cortex and cortically bound microtubule motors,[44] which enables spindle positioning relative to the mother-bud axis. In budding yeast, the bud neck and later the bud tip direct spindle positioning by providing a predetermined axis of cell division (Fig. 2A). In multicellular organisms, cells often use cues from surrounding cells, specifically from cell-cell adhesion sites, to orient the spindle and thus the plane of division (Fig. 2B).[38]

In *Drosophila* and vertebrates there are two *APC* genes: *Drosophila E-APC/dAPC2* and *dAPC1* and vertebrate *APC* and *APCL*.[17,45-48] EAPC/dAPC2 has been reported to localize to cell-cell junctions throughout the embryo, in the neuroepithelium and germ cells of the ovary and testis,[17] and this localization is actin dependent and coincides with Armadillo and DE-cadherin.[49] Thus, dAPC2 is in the correct location to play a role in spindle attachment at cell junctions and thereby orient the spindle and plane of division. In the syncytial blastoderm, *dAPC2* mutants lose nuclei from the cortex into the internal cytoplasm consistent with a loss of spindle attachment to the cortex.[18] Similarly, RNAi of *dAPC2* in the embryonic epidermis interferes with symmetric division along the planar axis of the embryo and instead these cells divide asymmetrically.[37] This suggests that dAPC2 is required for spindle orientation. Finally, in male germline stem cells (GSC), centrosomes are mispositioned in *dAPC2* mutants resulting in misoriented spindles (Fig. 2B).[38]

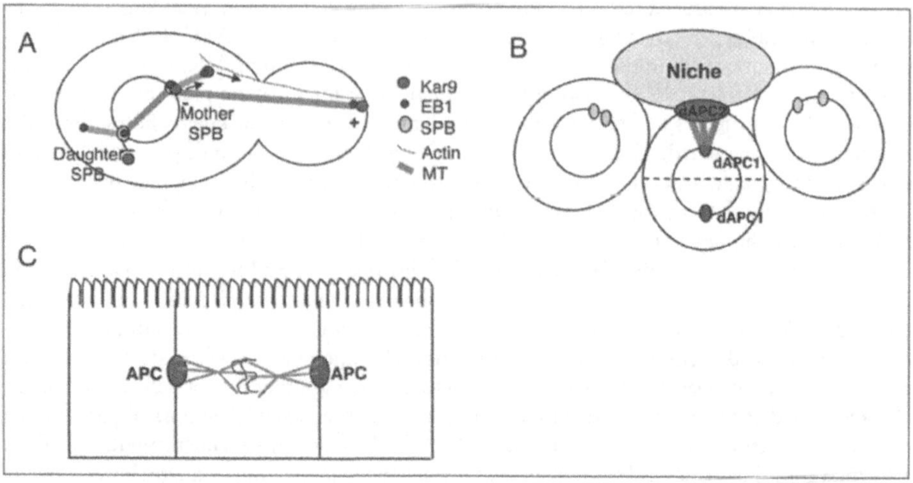

Figure 2. Potential roles for APC in spindle positioning in budding yeast and *Drosophila*. A) Model for Kar9-Bim1 mediated spindle positioning in budding yeast. Kar9 binds asymmetrically to the daughter-bound spindle pole body (Mother SPB) where it interacts with Bim1 (the yeast homologue of EB1) and is loaded onto the plus-end of a subset of microtubules polymerizing from the SPB.[40,41] Kar9-Bim1 bound microtubules are guided along actin cables via Myo2 (not shown) to the bud tip. Once attached to the bud tip, microtubule movement can be generated by microtubule plus-end dynamics against the cell cortex and cortically bound microtubule motors, which enables spindle positioning relative to the mother-bud axis. B) Model for dAPC2 and dAPC1 mediated spindle positioning in *Drosophila* male germline stem cells (GSC). dAPC2 localizes to the cell-cell junction between the stem cell and the hub cells where it may facilitate microtubule plus-end attachment.[38] The unattached centrosome migrates to the opposite pole to orient the spindle relative to the hub cells (dotted line marks plane of division). dAPC1 is also involved in spindle positioning, however dAPC1 localizes to the centrosomes, not the cell cortex.[38] C) In mammalian cells, tethering of astral microtubule plus-ends by APC at cell adhesion sites may orient the spindle in relation to the plane of the epithelium.

dAPC1 mutants also have defects in centrosome positioning in male GSCs, however only dAPC2 localizes to cortical attachment sites at cell junctions, whereas dAPC1 localizes to centrosomes in these cells[38] (see below and Fig. 2B). Thus, dAPC2 may play a key function in spindle orientation by providing cortical attachment sites at adherens junctions for microtubules (Fig. 2B), although the mechanism of attachment to adherens junctions by dAPC2 is not understood. One possibility is that dAPC2 functions in maintaining junctions and affects spindle positioning indirectly. However, fly embryos completely null for *dAPC1* and *dAPC2* do not display obvious defects in adhesion,[47] suggesting that dAPC2 may have a direct role in spindle positioning at cell junctions.

In vertebrates, APC depletion in mitotic cells results in loss of astral microtubules and mispositioning of the spindle relative to the geometric center of the cell.[39] APC localization to adherens junctions is weak and whether APC functions at adherens junctions in mammalian cells remains to be shown.[50-52] In mitotic MDCK cells, small clusters of APC can be found at the cell cortex, some of which align with astral microtubules (S.B., unpublished data). Thus, rather than tethering microtubules at adherens junctions, APC may provide cortical sites at the plasma membrane that stabilize and anchor the plus-ends of astral microtubules (Fig. 1C). Whether APC arrives at these sites independent of microtubules, or by traveling along microtubules and/or with microtubule plus-ends[25,53] remains to be shown. Furthermore, it will be important to determine whether these cortical sites also provide cues to orient the axis of the spindle in relation to the plane of the epithelium.

Disruption of APC's binding partner EB1 results in spindle mispositioining in *Drosophila*[37,54] and mammalian cells,[39] similar to the effects of disruption of APC. However, *Drosophila* dAPC2 lacks the vertebrate C-terminal EB1 and microtubule binding sites[17] and may not bind EB1 directly.[37] Therefore, it remains unclear whether dAPC2 and EB1 interact in a common pathway to regulate spindle orientation in *Drosophila.* In mammalian cells, depletion of both APC and EB1 does not have an additive affect, supporting the idea that APC and EB1 function in similar pathways governing spindle orientation.[39] One possibility is that, similar to yeast, APC capture of astral microtubule plus ends through EB1 assists spindle orientation in mammalian cells (Fig. 1C). Further experiments are needed to confirm that interactions between APC and EB1 in mammalian cells are required for spindle orientation.

Our knowledge about the function of APC in orienting the mammalian mitotic spindle is still at the early stages. Nevertheless, it is tempting to speculate that APC's role in positioning mitotic spindles is similar to its cytoskeletal roles at the cortex of interphase cells. In migrating cells, APC puncta localize to specialized cortical regions that promote microtubule growth.[13,25,53,55] At the basal cortex, APC puncta localize along the path of microtubules and provide points at which microtubules pause and rescue.[56,57] Thus, APC on the plasma membrane in mitotic cells may function similarly in guiding and stabilizing astral microtubules. Live cell imaging of mitotic cells to analyze the interaction of astral microtubule plus-ends with APC puncta is required to support this idea.

APC at Centrosomes: Potential Roles

APC localizes to mammalian centrosomes and this localization is conserved in the yeast analogue, Kar9 and the *Drosophila* homologue, dAPC1 (Fig. 2).[19,38,58,87] Centrosomes nucleate and anchor microtubules and are essential for establishment of a bipolar spindle.[59,60] APC mutant mouse embryonic stem cells (ES cells) have multipolar spindles and other centrosome abnormalities in mitosis indicating a role for APC at mitotic centrosomes.[15] Very little is known about the function of APC at mammalian centrosomes. However, data from lower eukaryotes suggest several interesting functions for APC at centrosomes including recruiting interacting proteins to this site.

In *S. cerevisiae,* Kar9 localizes asymmetrically to the daughter-bound spindle pole body (SPB), the yeast equivalent of the centrosome (Fig. 2A).[40,41,58] It is thought that the spindle pole body acts as a 'loading dock' for Kar9 assuring that only a subset of microtubules are loaded with the complex between Kar9 and Bim1.[41] As noted in the previous section, Kar9-Bim1 bound microtubules are then directed to the bud tip where Kar9 mediates attachment of microtubule plus-ends to the cortex (Fig. 2A).[34] In *Drosophila* testes, dAPC1 localizes to the centrosomes while dAPC2 localizes to the cortex (Fig. 2B).[38] Although the relationship between dAPC1 at centrosomes and dAPC2 at the cortex in *Drosophila* testes is unknown, mutants in either gene cause centrosome and spindle misorientation.[38] Vertebrate APC and EB1 localize specifically to the mother centriole, which is known to anchor a subset of microtubules.[19,60] Furthermore, APC decorates a subset of microtubules in migrating cells.[12] Thus, similar to yeast, a subset of microtubules anchored to the mother centriole may be loaded with APC and guided to their cortical destination (Fig. 1D).

In addition to APC, several other regulatory components of the Wnt signaling pathway localize to centrosomes. In the absence of a Wnt signal, a core destruction complex of APC, Axin and GSK3β control β-catenin levels by phosphorylating β-catenin[7,61] (see Kennell and Cadigan, this volume). Recent reports show that β-catenin also localizes to centrosomes in interphase and mitosis,[62-64] and regulation of its levels at centrosomes throughout the cell cycle may be important for proper centrosome function.[62,63] Overexpression of β-catenin results in increased centrosome number in interphase (SB unpublished results),[65] whereas β-catenin depletion inhibits centrosome separation in mitosis,[62,63] suggesting that there may be a threshold level of β-catenin that is important in some aspect of centrosome organization and function. Consistent with a requirement for regulated levels of β-catenin at centrosomes, GSK3β is active at centrosomes during interphase and is inactive at centrosomes in mitosis.[66] However, it is not known whether GSK3β activity at centrosomes regulates β-catenin levels at this location.

Phosphorylated β-catenin is recognized by the SCF (Skp1-cullin-F-box) ubiquitin ligase β-TrCP which ubiquitinates β-catenin marking it for degradation by the proteasome.[2,7] Components of the SCF ubiquitin ligase, Skp1 and Cul1, localize to centrosomes and are required for centriole duplication and separation in *Xenopus* egg extracts.[67] β-TrCP-/- mouse embryonic fibroblasts and *Drosophila* β-TrCP/Slimb mutants have overduplicated centrosomes.[68] Furthermore, components of the proteasome machinery have been localized to the centrosome and requirements for proteolysis in centrosome duplication and separation has been suggested.[69-71] Thus, most components of the core destruction complex and proteasome machinery are present and functional at centrosomes. A function for APC with the other components of the destruction complex at centrosomes may be in regulating β-catenin levels, which in turn could be important for normal centrosome duplication and separation.

APC localization to centrosomes together with several APC interacting proteins suggests several interesting possibilities for APC function at this site. APC, similar to Kar9 in yeast, may load onto a subset of microtubules in mitosis destined to perform specific functions such as cortical attachment, spindle orientation and chromosome congression. On the other hand, the centrosome may act as a signaling hub where Wnt signals coordinate with the cytoskeleton to perform specific tasks (see next section). Experiments that specifically inhibit APC at centrosomes will directly test for these potential functions.

Connecting APC Functions in Regulating Wnt Signaling and Microtubules

APC has divergent roles in cells: it is a regulator of Wnt signaling and also affects cytoskeletal function. In Wnt signaling, APC forms a complex that phosphorylates β-catenin leading to its degradation. As a cytoskeletal regulator, APC binds to proteins that either stabilize or destabilize microtubules at several locations in mitotic and interphase cells. Whether there is a functional link between Wnt signaling and regulation of microtubules through APC is unknown. One possibility is that APC can translate Wnt signals into cytoskeletal changes as it has been shown recently for a noncanonical Wnt signal during cell migration.[72] In the absence of Wnt signal, the pool of APC in the destruction complex is distinct from the pool of APC that binds to microtubules.[73] Moreover, GSK3β phosphorylation of APC inhibits the interaction of APC with microtubules,[74] supporting the idea that the pool of APC in the destruction complex may be unavailable to regulate microtubules. In the presence of a Wnt signal, GSK3β in the destruction complex is inhibited,[7] which may release APC from the destruction complex to interact with microtubules. Determining the influence of Wnt signaling on APC's role in microtubule regulation will provide a link between the function of APC in gene expression and microtubule dynamics and will provide insight into the synergistic consequences of *APC* mutations that give cells the clonal advantage needed for tumor formation (see below).

Multiple Mechansims by which *APC* Mutations Contribute to Cancer

Mutations in *APC* result in unregulated expression of Wnt-responsive genes and mitotic spindle defects. The mechanism by which *APC* mutations lead to unregulated gene expression and the consequences of this deregulation on cancer progression has been thoroughly examined.[7,75,76] Here, we will focus on potential consequences of *APC* mutations on centrosomes and kinetochores in mitosis, which could contribute to tumorigenesis by promoting chromosomal instability (see also Caldwell and Kaplan, this volume).

Most colorectal tumors exhibit chromosomal instability (CIN), which is caused by chromosome missegregation.[77-83] Abnormal centrosome number[84] and/or flawed attachment of microtubules to kinetochores[82] can lead to chromosome segregation defects. *APC* mutant mouse ES cells have multipolar spindles and abnormal centrosomes indicating that *APC* mutations could lead to defects in centrosomes.[15] Moreover, depletion of APC has been shown to cause lagging chromosomes, which correlates highly with reduced tension caused by lack of APC at kinetochores.[23] Thus, abnormal *APC* function at centrosomes or kinetochores could cause defects in mitosis that may result in chromosomal instability. It will be necessary to determine whether mutations in APC correlate with centrosome or chromosome abnormalities in colon cancer tissues. Since centrosome abnormalities could arise through indirect effects of abnormal mitoses,[85] it will be important to determine whether centrosome abnormalities are present in adenomas at the earliest stages of colon cancer.

A critical consequence of APC dysfunction in mitosis may be that in the absence of APC cells are able to bypass the mitotic checkpoint.[23,24] Depletion of APC results in a less efficient mitotic checkpoint,[23,24] which would allow cells with centrosome and spindle abnormalities to progress through mitosis with chromosome segregation defects.[28] Checkpoint proteins sense lack of tension on kinetochores and respond by inhibiting transition into anaphase.[28] After prolonged culture, colon cancer cells with *APC* mutations exhibit high levels of aneuploidy, indicating that problems in chromosome segregation bypass the checkpoint machinery.[86] Furthermore, depletion of APC does not cause an arrest in mitosis and cells progress through mitosis with lagging chromosomes, potentially a direct cause of aneuploidy.[23,24] In fact, APC depleted cells treated with low doses of nocodazole or taxol inappropriately exit from mitosis indicating a compromised mitotic checkpoint.[24] Reduced accumulation of Bub1 and BubR1 at kinetochores in APC depleted cells[24] might be the cause of premature mitotic exit allowing cells with kinetochore abnormalities to bypass the spindle checkpoint. Thus, APC function is required for two important kinetochore functions: proper microtubule attachment and regulation at kinetochores and activation of the mitotic checkpoint in response to chromosome missegregation. Mutations in APC may disrupt both these processes and promote CIN, thereby giving cells the clonal advantage needed for tumorigenesis.

In summary, disruption of the diverse functions of APC may synergistically contribute to cancer progression. Deregulated expression of Wnt-responsive genes promotes cell survival and proliferation, whereas defects in microtubule regulation at centrosomes and kinetochores could contribute to spindle abnormalities that are not properly recognized by the mitotic checkpoint and, therefore, cause CIN and, ultimately, cancer progression.

Acknowledgements

Work from the Nelson laboratory is supported by the NIH (GM35527) and a DOD predoctoral Fellowship to S.B. (BC050431).

References

1. Kinzler KW, Vogelstein B. Lessons from hereditary colorectal cancer. Cell 1996; 87(2):159-170.
2. Polakis P. The adenomatous polyposis coli (APC) tumor suppressor. Biochim Biophys Acta 1997; 1332(3):F127-147.
3. Kishida S, Yamamoto H, Ikeda S et al. Axin, a negative regulator of the wnt signaling pathway, directly interacts with adenomatous polyposis coli and regulates the stabilization of beta-catenin. J Biol Chem 1998; 273(18):10823-10826.
4. Rubinfeld B, Tice DA, Polakis P. Axin-dependent phosphorylation of the adenomatous polyposis coli protein mediated by casein kinase 1epsilon. J Biol Chem 2001; 276(42):39037-39045.
5. Sakanaka C, Weiss JB, Williams LT. Bridging of beta-catenin and glycogen synthase kinase-3-beta by axin and inhibition of beta-catenin-mediated transcription. Proc Natl Acad Sci USA 1998; 95(6):3020-3023.
6. Ikeda S, Kishida M, Matsuura Y et al. GSK-3beta-dependent phosphorylation of adenomatous polyposis coli gene product can be modulated by beta-catenin and protein phosphatase 2A complexed with Axin. Oncogene 2000; 19(4):537-545.
7. Polakis P. The oncogenic activation of beta-catenin. Curr Opin Genet Dev 1999; 9(1):15-21.
8. Munemitsu S, Souza B, Muller O et al. The APC gene product associates with microtubules in vivo and promotes their assembly in vitro. Cancer Res 1994; 54(14):3676-3681.
9. Su LK, Burrell M, Hill DE et al. APC binds to the novel protein EB1. Cancer Res 1995; 55(14):2972-2977.
10. Banks JD, Heald R. Adenomatous polyposis coli associates with the microtubule-destabilizing protein XMCAK. Curr Biol 2004; 14(22):2033-2038.
11. Jimbo T, Kawasaki Y, Koyama R et al. Identification of a link between the tumour suppressor APC and the kinesin superfamily. Nat Cell Biol 2002; 4(4):323-327.

12. Wen Y, Eng CH, Schmoranzer J et al. EB1 and APC bind to mDia to stabilize microtubules downstream of Rho and promote cell migration. Nat Cell Biol 2004; 6(9):820-830.
13. Nathke IS, Adams CL, Polakis P et al. The adenomatous polyposis coli tumor suppressor protein localizes to plasma membrane sites involved in active cell migration. J Cell Biol 1996; 134(1):165-179.
14. Zhou FQ, Zhou J, Dedhar S et al. NGF-induced axon growth is mediated by localized inactivation of GSK-3beta and functions of the microtubule plus end binding protein APC. Neuron 2004; 42(6):897-912.
15. Fodde R, Kuipers J, Rosenberg C et al. Mutations in the APC tumour suppressor gene cause chromosomal instability. Nat Cell Biol 2001; 3(4):433-438.
16. Kaplan KB, Burds AA, Swedlow JR et al. A role for the adenomatous polyposis coli protein in chromosome segregation. Nat Cell Biol 2001; 3(4):429-432.
17. McCartney BM, Dierick HA, Kirkpatrick C et al. Drosophila APC2 is a cytoskeletally-associated protein that regulates wingless signaling in the embryonic epidermis. J Cell Biol 1999; 146(6):1303-1318.
18. McCartney BM, McEwen DG, Grevengoed E et al. Drosophila APC2 and armadillo participate in tethering mitotic spindles to cortical actin. Nat Cell Biol 2001; 3(10):933-938.
19. Louie RK, Bahmanyar S, Siemers KA et al. Adenomatous polyposis coli and EB1 localize in close proximity of the mother centriole and EB1 is a functional component of centrosomes. J Cell Sci 2004; 117(Pt 7):1117-1128.
20. Biggins S, Walczak CE. Captivating capture: how microtubules attach to kinetochores. Curr Biol 2003; 13(11):R449-460.
21. Rieder CL, Salmon ED. The vertebrate cell kinetochore and its roles during mitosis. Trends Cell Biol 1998; 8(8):310-318.
22. Green RA, Kaplan KB. Chromosome instability in colorectal tumor cells is associated with defects in microtubule plus-end attachments caused by a dominant mutation in APC. J Cell Biol 2003; 163(5):949-961.
23. Draviam VM, Shapiro I, Aldridge B et al. Misorientation and reduced stretching of aligned sister kinetochores promote chromosome missegregation in EB1- or APC-depleted cells. EMBO J 2006; 25(12):2814-2827.
24. Dikovskaya D, Schiffmann D, Newton IP et al. Loss of APC induces polyploidy due to a combination of defects in mitosis and apoptosis. J Cell Biol, in press 2006.
25. Kita K, Wittmann T, Nathke IS et al. Adenomatous polyposis coli on microtubule plus ends in cell extensions can promote microtubule net growth with or without EB1. Mol Biol Cell 2006; 17(5):2331-2345.
26. Dikovskaya D, Newton IP, Nathke IS. The adenomatous polyposis coli protein is required for the formation of robust spindles formed in CSF Xenopus extracts. Mol Biol Cell 2004; 15(6):2978-2991.
27. Kroboth K, Newton IP, Kita K et al. Lack of adenomatous polyposis coli protein correlates with a decrease in cell migration and overall changes in microtubule stability. Mol Biol Cell 2007; 18(3):910-918.
28. Pinsky BA, Biggins S. The spindle checkpoint: tension versus attachment. Trends Cell Biol 2005; 15(9):486-493.
29. Dikovskaya D, Zumbrunn J, Penman GA et al. The adenomatous polyposis coli protein: in the limelight out at the edge. Trends Cell Biol 2001; 11(9):378-384.
30. Kinoshita K, Noetzel TL, Arnal I et al. Global and local control of microtubule destabilization promoted by a catastrophe kinesin MCAK/XKCM1. J Muscle Res Cell Motil 2006; 27(2):107-114.
31. Ahringer J. Control of cell polarity and mitotic spindle positioning in animal cells. Curr Opin Cell Biol 2003; 15(1):73-81.
32. Manneville JB, Etienne-Manneville S. Positioning centrosomes and spindle poles: looking at the periphery to find the centre. Biol Cell 2006; 98(9):557-565.
33. Beach DL, Thibodeaux J, Maddox P et al. The role of the proteins Kar9 and Myo2 in orienting the mitotic spindle of budding yeast. Curr Biol 2000; 10(23):1497-1506.
34. Bloom K. It's a kar9ochore to capture microtubules. Nat Cell Biol 2000; 2(6):E96-98.
35. Lee L, Tirnauer JS, Li J et al. Positioning of the mitotic spindle by a cortical-microtubule capture mechanism. Science 2000; 287(5461):2260-2262.
36. Korinek WS, Copeland MJ, Chaudhuri A et al. Molecular linkage underlying microtubule orientation toward cortical sites in yeast. Science. Mar 24 2000; 287(5461):2257-2259.
37. Lu B, Roegiers F, Jan LY et al. Adherens junctions inhibit asymmetric division in the drosophila epithelium. Nature 2001; 409(6819):522-525.
38. Yamashita YM, Jones DL, Fuller MT. Orientation of asymmetric stem cell division by the APC tumor suppressor and centrosome. Science 2003; 301(5639):1547-1550.
39. Green RA, Wollman R, Kaplan KB. APC and EB1 function together in mitosis to regulate spindle dynamics and chromosome alignment. Mol Biol Cell 2005; 16(10):4609-4622.

40. Paoletti A, Bornens M. Kar9 asymmetrical loading on spindle poles mediates proper spindle alignment in budding yeast. Dev Cell 2003; 4(3):289-290.
41. Liakopoulos D, Kusch J, Grava S et al. Asymmetric loading of Kar9 onto spindle poles and microtubules ensures proper spindle alignment. Cell 2003; 112(4):561-574.
42. Theesfeld CL, Irazoqui JE, Bloom K et al. The role of actin in spindle orientation changes during the Saccharomyces cerevisiae cell cycle. J Cell Biol 1999; 146(5):1019-1032.
43. Hwang E, Kusch J, Barral Y et al. Spindle orientation in saccharomyces cerevisiae depends on the transport of microtubule ends along polarized actin cables. J Cell Biol 2003; 161(3):483-488.
44. Pearson CG, Bloom K. Dynamic microtubules lead the way for spindle positioning. Nat Rev Mol Cell Biol 2004; 5(6):481-492.
45. Yu X, Waltzer L, Bienz M. A new drosophila APC homologue associated with adhesive zones of epithelial cells. Nat Cell Biol 1999; 1(3):144-151.
46. Nakagawa H, Murata Y, Koyama K et al. Identification of a brain-specific APC homologue, APCL and its interaction with beta-catenin. Cancer Res 1998; 58(22):5176-5181.
47. McCartney BM, Price MH, Webb RL et al. Testing hypotheses for the functions of APC family proteins using null and truncation alleles in drosophila. Development 2006; 133(12):2407-2418.
48. van Es JH, Kirkpatrick C, van de Wetering M et al. Identification of APC2, a homologue of the adenomatous polyposis coli tumour suppressor. Curr Biol 1999; 9(2):105-108.
49. Mimori-Kiyosue Y, Tsukita S. Where is APC going? J Cell Biol 2001; 154(6):1105-1109.
50. Barth AI, Nathke IS, Nelson WJ. Cadherins, catenins and APC protein: Interplay between cytoskeletal complexes and signaling pathways. Curr Opin Cell Biol 1997; 9(5):683-690.
51. Bienz M, Hamada F. Adenomatous polyposis coli proteins and cell adhesion. Curr Opin Cell Biol 2004; 16(5):528-535.
52. Barth AI, Nelson WJ. What can humans learn from flies about adenomatous polyposis coli? Bioessays 2002; 24(9):771-774.
53. Mimori-Kiyosue Y, Shiina N, Tsukita S. Adenomatous polyposis coli (APC) protein moves along microtubules and concentrates at their growing ends in epithelial cells. J Cell Biol 2000; 148(3):505-518.
54. Rogers SL, Rogers GC, Sharp DJ et al. Drosophila EB1 is important for proper assembly, dynamics and positioning of the mitotic spindle. J Cell Biol 2002; 158(5):873-884.
55. Mimori-Kiyosue Y, Shiina N, Tsukita S. The dynamic behavior of the APC-binding protein EB1 on the distal ends of microtubules. Curr Biol 2000; 10(14):865-868.
56. Reilein A, Nelson WJ. APC is a component of an organizing template for cortical microtubule networks. Nat Cell Biol 2005; 7(5):463-473.
57. Reilein A, Yamada S, Nelson WJ. Self-organization of an acentrosomal microtubule network at the basal cortex of polarized epithelial cells. J Cell Biol 2005; 171(5):845-855.
58. Miller RK, Rose MD. Kar9p is a novel cortical protein required for cytoplasmic microtubule orientation in yeast. J Cell Biol 1998; 140(2):377-390.
59. Murphy SM, Stearns T. Cytoskeleton: Microtubule nucleation takes shape. Curr Biol 1996; 6(6):642-644.
60. Bornens M. Centrosome composition and microtubule anchoring mechanisms. Curr Opin Cell Biol 2002; 14(1):25-34.
61. Peifer M, Polakis P. Wnt signaling in oncogenesis and embryogenesis—A look outside the nucleus. Science 2000; 287(5458):1606-1609.
62. Kaplan DD, Meigs TE, Kelly P et al. Identification of a role for beta-catenin in the establishment of a bipolar mitotic spindle. J Biol Chem 2004; 279(12):10829-10832.
63. Bahmanyar S, Kaplan DD, DeLuca JG et al. B-catenin is a Nek2 substrate involved in centrosome separation. Genes Dev 2006; 22(1):91-105.
64. Huang P, Senga T, Hamaguchi M. A novel role of phospho-beta-catenin in microtubule regrowth at centrosome. Oncogene 2007; 26(30):4357-4371.
65. Ligon LA, Karki S, Tokito M et al. Dynein binds to beta-catenin and may tether microtubules at adherens junctions. Nat Cell Biol 2001; 3(10):913-917.
66. Wakefield JG, Stephens DJ, Tavare JM. A role for glycogen synthase kinase-3 in mitotic spindle dynamics and chromosome alignment. J Cell Sci 2003; 116(Pt 4):637-646.
67. Freed E, Lacey KR, Huie P et al. Components of an SCF ubiquitin ligase localize to the centrosome and regulate the centrosome duplication cycle. Genes Dev 1999; 13(17):2242-2257.
68. Wojcik EJ, Glover DM, Hays TS. The SCF ubiquitin ligase protein slimb regulates centrosome duplication in Drosophila. Curr Biol 2000; 10(18):1131-1134.
69. Fabunmi RP, Wigley WC, Thomas PJ et al. Activity and regulation of the centrosome-associated proteasome. J Biol Chem 2000; 275(1):409-413.
70. McDonald HB, Byers B. A proteasome cap subunit required for spindle pole body duplication in yeast. J Cell Biol 1997; 137(3):539-553.

71. Winey M, Baum P, Goetsch L et al. Genetic determinants of spindle pole body duplication in budding yeast. Cold Spring Harb Symp Quant Biol 1991; 56:705-708.
72. Schlessinger K ME, Hall A. Cdc42 and noncanonical Wnt signal transduction pathways cooperate to promote cell polarity. J Cell Biol 2007; 178(3):355-361.
73. Penman GA, Leung L, Nathke IS. The adenomatous polyposis coli protein (APC) exists in two distinct soluble complexes with different functions. J Cell Sci 2005; 118(Pt 20):4741-4750.
74. Zumbrunn J, Kinoshita K, Hyman AA et al. Binding of the adenomatous polyposis coli protein to microtubules increases microtubule stability and is regulated by GSK3 beta phosphorylation. Curr Biol 2001; 11(1):44-49.
75. Bienz M, Clevers H. Linking colorectal cancer to Wnt signaling. Cell 2000; 103(2):311-320.
76. Michor F, Iwasa Y, Lengauer C et al. Dynamics of colorectal cancer. Semin Cancer Biol 2005; 15(6):484-493.
77. Lengauer C, Kinzler KW, Vogelstein B. Genetic instability in colorectal cancers. Nature 1997; 386(6625):623-627.
78. Lengauer C, Kinzler KW, Vogelstein B. Genetic instabilities in human cancers. Nature 1998; 396(6712):643-649.
79. Nowak MA, Komarova NL, Sengupta A et al. The role of chromosomal instability in tumor initiation. Proc Natl Acad Sci USA 2002; 99(25):16226-16231.
80. Rajagopalan H, Nowak MA, Vogelstein B et al. The significance of unstable chromosomes in colorectal cancer. Nat Rev Cancer 2003; 3(9):695-701.
81. Giaretti W, Venesio T, Prevosto C et al. Chromosomal instability and APC gene mutations in human sporadic colorectal adenomas. J Pathol 2004; 204(2):193-199.
82. Draviam VM, Xie S, Sorger PK. Chromosome segregation and genomic stability. Curr Opin Genet Dev 2004; 14(2):120-125.
83. Michor F, Iwasa Y, Vogelstein B et al. Can chromosomal instability initiate tumorigenesis? Semin Cancer Biol 2005; 15(1):43-49.
84. Brinkley BR, Goepfert TM. Supernumerary centrosomes and cancer: Boveri's hypothesis resurrected. Cell Motil Cytoskeleton 1998; 41(4):281-288.
85. Nigg EA. Centrosome aberrations: Cause or consequence of cancer progression? Nat Rev Cancer 2002; 2(11):815-825.
86. Tighe A, Johnson VL, Albertella M et al. Aneuploid colon cancer cells have a robust spindle checkpoint. EMBO Rep 2001; 2(7):609-614.
87. Akong, K, Grevengoed EE, Price MH et al. Drosophila APC2 and APC1 play overlapping roles in wingless signaling in the embryo and imaginal discs. Development 2002; 250:91-100.

Chapter 7

The Adenomatous Polyposis Coli Tumor Suppressor and Wnt Signaling in the Regulation of Apoptosis

Hassina Benchabane and Yashi Ahmed*

Abstract

The adenomatous polyposis coli (APC) tumor suppressor is an essential negative regulator in the evolutionarily conserved Wnt/Wingless (Wg) signal transduction pathway. During normal development, Wnt signaling is required not only to induce cell proliferation and cell fate specification, but also to induce apoptotic cell death. However in some malignant states triggered by APC loss, inappropriate activation of Wnt signaling promotes cell survival and inhibits cell death, indicating that the cellular response to APC loss and Wnt signaling is highly dependent on cell context. This chapter summarizes our current understanding of the role of APC and Wnt signaling in the regulation of apoptosis, based upon studies from fly and mouse in vivo models, as well as cultured carcinoma cells.

Introduction

Mutation of both alleles of the human *adenomatous polyposis coli* (*APC*) tumor suppressor gene triggers the development of upper and lower gastrointestinal polyps and carcinoma and less frequently, hepatocellular carcinoma and hepatoblastoma.[1-7] In addition to developing a number of extra-intestinal manifestations, individuals with a germline mutation in one *APC* allele develop hundreds to thousands of neoplastic colorectal polyps as a consequence of somatic mutation in the other *APC* allele (reviewed in ref. 8). Some of these polyps invariably progress to carcinoma and without surgical intervention, the mean age at diagnosis of colorectal adenocarcinoma is 39 years.[9] Biallelic inactivation of *APC* is also found in the earliest developmental stages of greater than 80% of sporadic colorectal carcinomas, which are the second and third leading cause of cancer-related death, respectively, in men and women living in the United States.[10-12] Together, these observations identify *APC* as a primary gatekeeper of cell proliferation and survival in the colonic epithelium.[13] (see also Kwong and Dove, this volume)

APC is an essential negative regulator in the evolutionarily conserved Wnt/Wingless (Wg) signal transduction pathway (Fig. 1).[14-16] In the absence of Wnt signaling, APC forms a "destruction complex" in the cytoplasm with Glycogen Synthase Kinase 3 (GSK3), Casein Kinase 1 (CK1), Axin and β-catenin/Armadillo (Arm) (see also Kennell and Cadigan, this volume). GSK3 and CK1 sequentially phosphorylate specific serine and threonine residues in the amino-terminus of β-catenin/Arm, an event that depends on the scaffolding functions of Axin and APC. Phosphorylated β-catenin/Arm is targeted for ubiquitination by the E3 ubiquitin-ligase β-TrCP and subsequently degraded by the proteasome. Binding of the Wnt ligand to its coreceptors Frizzled

*Corresponding Author: Yashi Ahmed—Dartmouth Medical School, Department of Genetics, Hanover, NH, USA. Email: yashi.f.ahmed@dartmouth.edu

APC Proteins, edited by Inke S. Näthke and Brooke M. McCartney. ©2009 Landes Bioscience and Springer Science+Business Media.

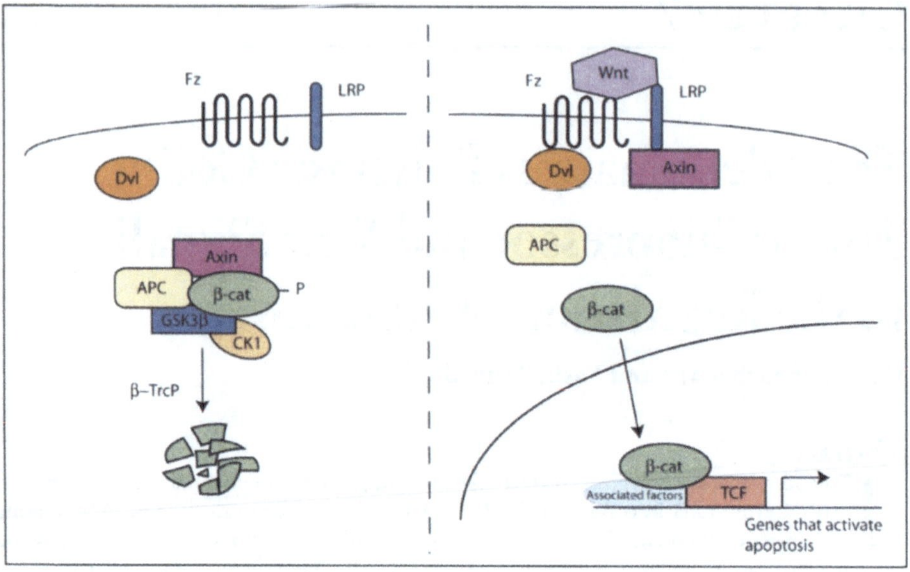

Figure 1. Schematic representation of the evolutionarily conserved Wnt/Wingless signal transduction pathway.

and Low-density lipoprotein-receptor-related proteins 5/6 (LRP5/6), stimulates the dissociation of Axin from APC and leads to the stabilization of cytoplasmic and nuclear β-catenin/Arm. Nuclear β-catenin/Arm binds to the transcription factor T-cell factor/lymphoid enhancement factor (TCF/LEF) and activates transcription of target genes in cooperation with Legless/Bcl-9 and Pygopus. β-catenin/Arm regulates the transcription of genes involved in a number of cellular events, including cell fate determination, cell proliferation, cell differentiation and apoptosis.

This chapter summarizes our current understanding of the role of APC and Wnt signaling in induction of apoptotic cell death. Activation of the Wnt signaling pathway has been shown to both positively and negatively regulate apoptosis.[17-27] In concordance with its essential role as a negative regulator of Wnt signaling, APC activity impacts the induction of programmed cell death in a number of cell types. We describe examples from fly and mouse model systems in which APC is required to prevent apoptosis and thereby promote cell survival during normal development. Conversely, analysis of APC loss in cultured colonic carcinoma cells reveals that the resultant activation of β-catenin signaling promotes cell survival and inhibits cell death. Together, these studies indicate an essential role for APC and Wnt signaling in regulating the decision to undergo apoptotic death and reveal the importance of cell context in determining the response to APC loss.

Overview of Apoptotic Cell Death

Apoptotic cell death is an evolutionarily-conserved process that is critical for normal development and for homeostasis in the adult life of animals (reviewed in ref. 28). During embryogenesis, apoptotic death is required for the patterning of many tissues and for eliminating tissues that have outlived their usefulness. In adults, apoptosis serves to eliminate infected cells, cells having undergone DNA damage and inappropriately proliferating cells. In addition, apoptotic death is also important in selecting the immune repertoire and for maintaining homeostasis in tissue size. Apoptosis is particularly important in maintaining homeostasis in regenerating tissues, such as the absorptive epithelium of the small intestine, which in mice, turns over entirely every three to five days (reviewed in ref. 14). Maintenance of the normal crypt-villus intestinal structures requires continual production of new cells in the crypt compartment, as well as elimination of cells by

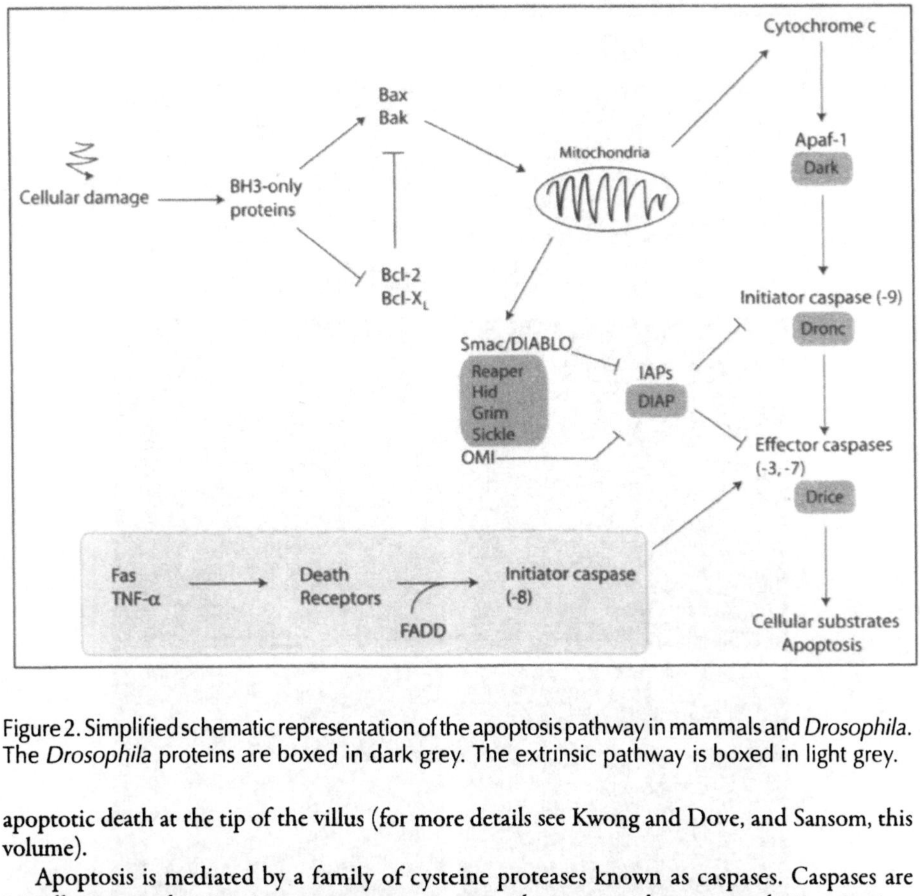

Figure 2. Simplified schematic representation of the apoptosis pathway in mammals and *Drosophila*. The *Drosophila* proteins are boxed in dark grey. The extrinsic pathway is boxed in light grey.

apoptotic death at the tip of the villus (for more details see Kwong and Dove, and Sansom, this volume).

Apoptosis is mediated by a family of cysteine proteases known as caspases. Caspases are initially expressed as inactive pro-caspase precursors that must undergo proteolytic activation. Initiator caspases are activated in response to developmental signals and cellular stress and cleave the precursor forms of effector caspases. Activated effector caspases in turn cleave a series of cellular substrates, such as Bid, Retinoblastoma and Poly(ADP-ribose) polymerase (reviewed in ref. 29), resulting in apoptosis. Two pathways trigger caspase activation and apoptosis in mammalian cells: the intrinsic and extrinsic pathways (Fig. 2; reviewed in refs. 30-32). The intrinsic apoptosis pathway involves the release of Cytochrome c from mitochondria and is regulated through the Bcl-2 family of proteins, while the extrinsic apoptosis pathway is initiated by the binding of death ligands, such as Fas and TNF-α, to cell surface death receptors.[30,31]

APC Prevents Neuronal Apoptosis during Retinal Development in *Drosophila*

As the Wnt/Wg signaling pathway is evolutionarily conserved, genetic studies in *Drosophila* have been instrumental in identifying new components in the pathway and for elucidating their in vivo function. While Wg signaling has well-established roles in cell proliferation and cell fate specification during normal development, recent studies have revealed that Wg signaling is also important for developmentally regulated apoptotic cell death and highlight the importance of β-catenin/Arm signaling in this process. Two well-established examples of Wg-induced apoptosis occur during the development of the fly retina. First, Wg signaling is required to refine the highly-ordered compound eye structure by eliminating excess cells that are present between each of approximately 750 unit ommatidia.[33] Second, Wg signaling is also required to induce the death

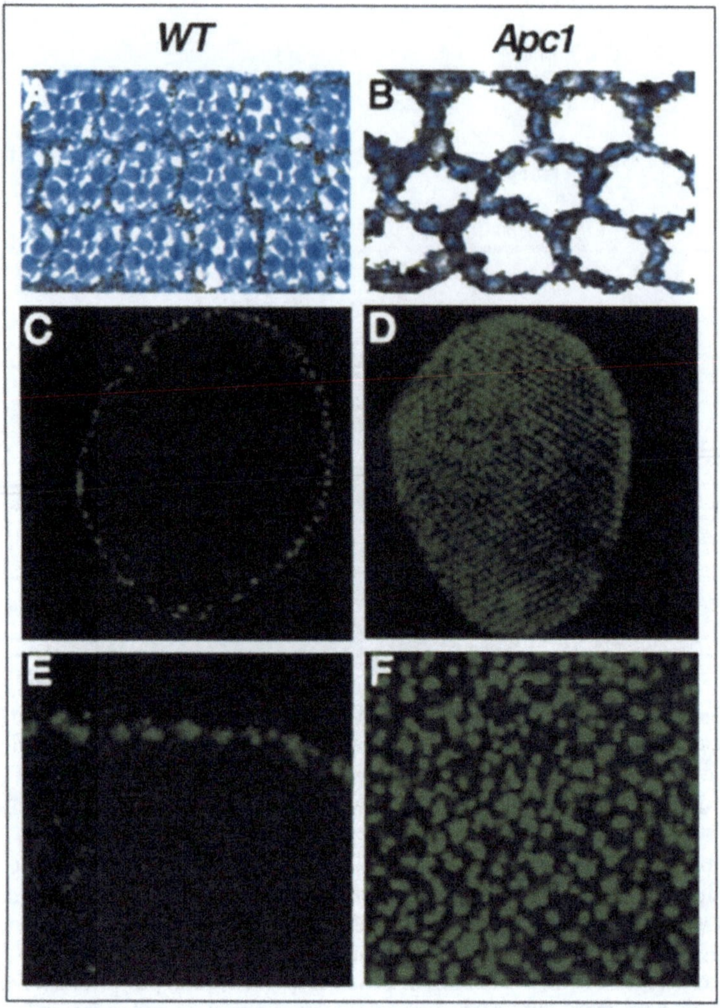

Figure 3. Inactivation of *Apc1* causes apoptosis of *Drosophila* retinal photoreceptors. (A, C, E) Wild type (WT) eye. (B, D, F) *Apc1* mutant. Cross-sections of the retina are presented in A and B. In WT (A), Photoreceptors (in blue) are surrounded by pigment cells. In the *Apc1* mutant (B), photoreceptors are absent and only pigment cells remain. Activated Caspase-3 expression is presented in C-F. Whole retinas are shown in C and D and close-ups of the retina are shown in E and F.

of all retinal neurons (photoreceptors) at the edge of the eye (Fig. 3), a process believed to be important for eliminating defective neurons and for sculpting of the retinal periphery.[18,34] During eye development, Wg is secreted by epithelial cells that surround the retina and spreads into the eye edge to activate β-catenin/Arm signaling. Transduction of high-level β-catenin/Arm signaling activates the transcription of three major apoptosis effectors, *reaper, hid* and *grim* (see Fig. 2). Inactivation of Wg signaling in the retinal periphery results in an excessive accumulation of retinal neurons and a disordered eye structure, revealing an essential role for Wg in developmentally regulated apoptotic death.[18,34]

Analysis of APC function in the fly eye confirms the critical role of Wg signaling in regulating apoptosis during retinal development. In *Drosophila*, as in humans, there are two APC homologs (*Apc1* and *Apc2*) and genetic studies have provided evidence that the function of APC as a critical negative regulator of β-catenin signaling is conserved from flies to humans.[17,35,36] Inactivation of *Apc1* results in hyperactivation of β-catenin/Arm signaling in all retinal photoreceptors,[17,37] which induces both their apoptotic death and a concomitant hypertrophy of the neighboring retinal pigment cells (Fig. 3).[17] Supporting the conclusion that this photoreceptor apoptosis results from ectopic β-catenin/Arm signaling, the apoptosis can be suppressed simply by reducing the gene dosage of *Arm*, or its transcriptional activator *dTCF/Pangolin*. Conversely, in otherwise wild-type flies, photoreceptor apoptosis can be induced by overexpression of Wingless or Arm.[17,38,39] Thus, the apoptotic response of all photoreceptors upon Apc1 loss is similar to the response of peripheral photoreceptors to Wg signaling at the retinal edge during normal eye development.[18,34] Furthermore, in both wild-type and in *Apc1* mutant flies, elimination of the apoptosis effectors *hid*, *reaper* and *grim* prevents photoreceptor apoptosis;[18] however, whether these three genes are direct targets of Wg transduction remains to be elucidated. Together, these studies reveal a critical role for Wg transduction in promoting apoptosis and for APC in promoting cell survival in retinal cells in which Wg is low or absent.

Unexpectedly, the fly retinal cell death phenotype induced by *Apc1* loss parallels a retinal defect found in humans with germline mutations in *APC*, termed congenital hypertrophy of the retinal pigment epithelium.[40] In these individuals, congenital retinal lesions are often bilateral and multifocal. In the rare instances in which these lesions have been sectioned at autopsy, they have been found to be composed of degenerated photoreceptors and hypertrophied pigment cells.[41,42] Thus, reduction of APC activity in flies and humans results in retinal defects that are at least superficially similar. A mouse model for APC loss recapitulates these retinal lesions[43] and provides a means to determine whether, as in the fly, the retinal neuronal degeneration induced by APC loss in mammals results from an apoptotic response to hyperactivation of β-catenin signaling. This model will also be useful in determining whether heterozygosity for *APC* is sufficient to induce these changes, or whether, like in the intestinal epithelium, both wild type copies have to be lost.

APC Prevents the Apoptotic Death of Mammalian Cephalic and Cardiac Neural Crest Cells

Neural crest development in the mouse provides another interesting example of the role of APC in preventing apoptosis and promoting cell survival. Neural crest cells are a multipotent stem cell population that migrates from the dorsal neural tube to diverse positions throughout the body, differentiating into a variety of cell types.[44] Derivatives of neural crest include bone and cartilage tissues, peripheral nerves, glia, smooth muscle cells, Schwann cells and melanocytes. The Wnt signaling pathway plays an important role in the early stages of neural crest development, such as neural crest induction and melanocyte formation.[45-48] Hasegawa et al[49] analyzed the function of APC in neural crest cells by specifically inactivating *Apc* in the neural crest of mice, using the Cre-loxP recombination system. The mutant mice generated had markedly increased apoptosis, as revealed by TUNEL staining, in the ventral craniofacial mesenchyme, the branchial arch and the cardiac outflow tract, indicating that APC loss leads to the apoptosis of a subset of neural crest derivatives. As a result, these mice had severe craniofacial and cardiac defects and died shortly after birth. All bones derived from the cephalic neural crest were affected and cardiac defects included ventricular septal defects and persistent truncus arteriosus. Remarkably, neural crest derivatives destined to become bone or cartilage undergo apoptosis, wheras those that differentiate as peripheral nerves, Schwann cells, or melanocytes survive, indicating that the apoptotic response of neural crest cells to APC loss is context dependent. Increased β-catenin levels were observed in tissues containing TUNEL-positive cells, indicating that the apoptosis induced by APC loss may result from increased β-catenin activity; however a direct link between the apoptosis and increased Wnt signaling awaits further investigation.

APC Loss and Activation of Wnt Signaling Results in Both Increased Cell Proliferation and Increased Apoptosis in Mammalian Intestinal Epithelia

Analysis of the mammalian intestinal epithelium exposed to different levels of Wnt signaling revealed unexpected effects on proliferation and apoptosis. The mammalian intestinal epithelium is a self-renewing tissue organized into highly ordered structures composed of villi and crypts (see Kwong and Dove, and Sansom, this volume). Mitotically active stem cells present at the base of each crypt migrate along the crypt-villus axis where they differentiate, carry out their specific role in the epithelium and finally undergo apoptosis and are shed into the gut lumen. The Wnt signaling pathway is a key regulator of cell fate along the crypt-villus axis.[14,50-52] Inactivation of Wnt signaling results in a complete loss of the crypt progenitor compartment, while hyperactivation of signaling results in ectopic expression of Wnt-target genes and hyperproliferation of cells in the crypt compartment. As a result of Wnt pathway activation, epithelial cells displayed a number of phenotypes associated with colorectal lesions. Migration of epithelial cells along the crypt-villus axis was abrogated, differentiation was arrested and increased proliferation was observed. In addition, proliferation was no longer confined to the base of the crypt. Together, these studies reveal the critical role of Wnt signaling in regulating self-renewal of the intestinal epithelium.

Unexpectedly, hyperactivation of Wnt signaling in the intestinal epithelium results not only in increased cell proliferation, but also in increased apoptosis. Increased apoptosis was observed in three distinct mouse models that address the effects of increased Wnt signaling in the intestinal epithelium. First, directed expression of a constitutively activated β-catenin, similar to that found in some colonic carcinomas, resulted not only in increased cell proliferation in crypt epithelial cells, but also in increased apoptotic death.[53] Second, conditional inactivation of APC in the mouse intestinal epithelium led to the accumulation and nuclear translocation of β-catenin and increased Wnt signaling.[54] This in vivo model for increased Wnt signaling resulted in qualitatively similar results: not only was cell proliferation increased, but apoptotic cell death was increased also. Third, expression of a β-catenin/Lef-1 fusion protein that enhances Wnt signaling in the intestinal epithelium of a chimeric mouse model also resulted in increased apoptosis.[55] In these mice, increased apoptosis was restricted to only those intestinal epithelial cells that expressed the β-catenin/Lef-1 fusion protein and unexpectedly was not associated with increased cell proliferation.

Based on these data, it has been proposed that different levels of β-catenin signaling result in qualitatively distinct cellular responses.[55] In this model, high levels of β-catenin signaling in the intestinal epithelia induce apoptosis, wheras intermediate levels of signaling result in sustained cell proliferation. Conversely, complete loss of β-catenin signaling results in the absence of proliferation and differentiation. Evidence supporting dosage-dependent cellular responses to APC loss and β-catenin signaling has been documented in several models, including mouse embryonic stem cells.[56,57] Whether high-level β-catenin signaling is important for an apoptotic response in intestinal epithelia and whether there exists a direct link between β-catenin mediated target gene activation and the induction of apoptosis in these cells awaits further experimental analysis.

Promotion of Cell Survival and Negative Regulation of Apoptosis by Wnt Signaling in Carcinomas

Although activation of Wnt signaling can induce both proliferation and apoptosis during normal development, Wnt signaling is also thought to have the opposite role of promoting cell survival and increasing cell proliferation in cancer cells. For instance, Wnt expression is upregulated in a number of human cancers[58-64] and monoclonal antibodies or siRNA directed against Wnt-1 and Wnt-2 in cultured carcinoma cells leads to β-catenin downregulation and induction of apoptosis.[22-25] When either melanoma, non-small cell lung carcinoma, breast carcinoma, mesothelioma, or sarcoma cells overexpressing Wnt-1 or Wnt-2 are injected in nude mice along with the Wnt-1 or Wnt-2 monoclonal antibody, respectively, tumor growth is inhibited and more apoptotic cells are observed compared to mice injected with the Wnt-1 or Wnt-2 overexpressing cells alone.[22,24]

Similarly, in several different types of cultured carcinoma cells, decreased β-catenin levels induced by forced expression of members of the destruction complex lead to induction of apoptosis. Hepatocellular and colorectal carcinoma cells, for instance, have high levels of nuclear β-catenin due to loss of function mutations in *APC* or *Axin*, or hyperactivating mutations in *β-catenin*. Expressing APC or Axin in these cells induces apoptosis.[65] Similarly, expressing APC or a dominant negative TCF in a colon carcinoma cell line lacking endogenous APC increases caspase expression and activity and stimulates apoptosis.[66,67] Together, these studies provide evidence that activation of Wnt signaling is important for promoting cell survival in some types of carcinoma cells and again suggest that the cellular response to Wnt signaling is highly dependent upon cell context.

Recent studies have identified several candidate target genes that prevent apoptosis and are regulated by Wnt signaling. Huang et al[68] analyzed the expression profile of HeLa cells in which β-catenin expression was downregulated by RNAi. The expression of a number of apoptosis-related genes, including *MYBL2, BAG2, BAG3, PTEN, HIF1A, PDCD6IP* and *DAP3,* is increased in these cells. The anti-apoptotic protein Bcl-X_L has also been identified as a potential target of Wnt transduction.[69] In thymocytes, whose survival depends on Wnt signaling, inactivation of TCF-1 leads to a decrease in Bcl-X_L expression. These studies are among the first to delineate targets of Wnt transduction that are important for regulating apoptosis. It is important to note that disruption of APC may affect apoptosis independently of Wnt signaling as well.[70,71] Dikovskaya et al recently demonstrated that loss of APC resulted in decreased apoptosis to the same degree in wild type cultured cells as in a cell line where the Wnt pathway was activated by mutation in β-catenin.[70] The mechanisms by which loss of APC may affect apoptosis independent of Wnt signaling are not known.

Conclusion

The Wnt signal transduction pathway has well-established roles in promoting cell proliferation and differentiation in both vertebrates and invertebrates. Recent studies have revealed that during normal development, Wnt signaling also has an essential role in promoting apoptosis. In some cells that are not normally destined to die, loss of APC activates ectopic Wnt signaling and results in excessive apoptotic death, revealing that APC is required to promote cell survival in different developmental contexts. Conversely, in some cultured carcinoma cells, Wnt signaling is critical for cell survival and inhibition of apoptosis and recent work has revealed a number of putative target genes that might directly regulate this Wnt signaling-induced response. Together, these studies in both in vivo models and cultured cell lines highlight the importance of cellular context in determining the apoptotic response to APC loss and Wnt transduction.

How do some cells respond to the loss of APC by proliferating, while others respond by dying? One current challenge in addressing this question is to determine the cell contexts in which elevated β-catenin signaling directly induces an apoptotic response upon APC loss and whether apoptosis results specifically from high-level of β-catenin signaling. The recent generation of numerous reagents that allow either conditional activation or conditional inhibition of Wnt transduction will be instrumental in addressing this issue. In addition, determining the molecular factors that establish the context of a cell and thereby influence the cellular response to activated Wnt signaling, also awaits further analysis. Understanding how these effectors regulate the decision to induce proliferation, differentiation, or death is of fundamental importance in developing novel therapies that redirect APC mutant colonic tumor cells from a proliferation to a cell death program.

Acknowledgements

We thank Edward G. Hughes and Jason R. Baird for the photographs in Figure 3. Work in our laboratory is supported by the Norris Cotton Cancer Center, the Emerald Foundation, the Scholars Program of the General Motors Cancer Research Foundation, the American Cancer Society (IRG-82-003-21), the National Institutes of Health (RO1CA105038) and the Howard Hughes Medical Institute, through an award from the Biomedical Research Support Program for Medical Schools to Dartmouth Medical School (76200-560801).

References

1. Groden J, Thliveris A, Samowitz W et al. Identification and characterization of the familial adenomatous polyposis coli gene. Cell 1991; 66(3):589-600.
2. Joslyn G, Carlson M, Thliveris A et al. Identification of deletion mutations and three new genes at the familial polyposis locus. Cell 1991; 66(3):601-613.
3. Kinzler KW, Nilbert MC, Su LK et al. Identification of FAP locus genes from chromosome 5q21. Science 1991; 253(5020):661-665.
4. Nishisho I, Nakamura Y, Miyoshi Y et al. Mutations of chromosome 5q21 genes in FAP and colorectal cancer patients. Science 1991; 253(5020):665-669.
5. Gruner BA, DeNapoli TS, Andrews W et al. Hepatocellular carcinoma in children associated with Gardner syndrome or familial adenomatous polyposis. J Pediatr Hematol Oncol 1998; 20(3):274-278.
6. Attard TM, Cuffari C, Tajouri T et al. Multicenter experience with upper gastrointestinal polyps in pediatric patients with familial adenomatous polyposis. Am J Gastroenterol 2004; 99(4):681-686.
7. Hirschman BA, Pollock BH, Tomlinson GE. The spectrum of APC mutations in children with hepatoblastoma from familial adenomatous polyposis kindreds. J Pediatr 2005; 147(2):263-266.
8. Bronner MP. Gastrointestinal polyposis syndromes. Am J Med Genet A 2003; 122(4):335-341.
9. Giardiello FM, Brensinger JD, Petersen GM. AGA technical review on hereditary colorectal cancer and genetic testing. Gastroenterology 2001; 121(1):198-213.
10. Miyoshi Y, Ando H, Nagase H et al. Germ-line mutations of the APC gene in 53 familial adenomatous polyposis patients. Proc Natl Acad Sci USA 1992; 89(10):4452-4456.
11. Powell SM, Petersen GM, Krush AJ et al. Molecular diagnosis of familial adenomatous polyposis. N Engl J Med 1993; 329(27):1982-1987.
12. Smith KJ, Johnson KA, Bryan TM et al. The APC gene product in normal and tumor cells. Proc Natl Acad Sci USA 1993; 90(7):2846-2850.
13. Kinzler KW, Vogelstein B. Lessons from hereditary colorectal cancer. Cell 1996; 87(2):159-170.
14. Clevers H. Wnt/beta-Catenin Signaling in Development and Disease. Cell 2006; 127(3):469-480.
15. Willert K, Jones KA. Wnt signaling: is the party in the nucleus? Genes Dev 2006; 20(11):1394-1404.
16. Stadeli R, Hoffmans R, Basler K. Transcription under the control of nuclear Arm/beta-catenin. Curr Biol 2006; 16(10):R378-385.
17. Ahmed Y, Hayashi S, Levine A et al. Regulation of armadillo by a Drosophila APC inhibits neuronal apoptosis during retinal development. Cell 1998; 93(7):1171-1182.
18. Lin HV, Rogulja A, Cadigan KM. Wingless eliminates ommatidia from the edge of the developing eye through activation of apoptosis. Development 2004; 131(10):2409-2418.
19. Olmeda D, Castel S, Vilaro S et al. Beta-catenin regulation during the cell cycle: implications in G2/M and apoptosis. Mol Biol Cell 2003; 14(7):2844-2860.
20. Kim K, Pang KM, Evans M et al. Overexpression of beta-catenin induces apoptosis independent of its transactivation function with LEF-1 or the involvement of major G1 cell cycle regulators. Mol Biol Cell 2000; 11(10):3509-3523.
21. Edlund S, Lee SY, Grimsby S et al. Interaction between Smad7 and beta-catenin: importance for transforming growth factor beta-induced apoptosis. Mol Cell Biol 2005; 25(4):1475-1488.
22. He B, You L, Uematsu K et al. A monoclonal antibody against Wnt-1 induces apoptosis in human cancer cells. Neoplasia 2004; 6(1):7-14.
23. You L, He B, Uematsu K et al. Inhibition of Wnt-1 signaling induces apoptosis in beta-catenin-deficient mesothelioma cells. Cancer Res 2004; 64(10):3474-3478.
24. You L, He B, Xu Z et al. An anti-Wnt-2 monoclonal antibody induces apoptosis in malignant melanoma cells and inhibits tumor growth. Cancer Res 2004; 64(15):5385-5389.
25. You L, He B, Xu Z et al. Inhibition of Wnt-2-mediated signaling induces programmed cell death in nonsmall-cell lung cancer cells. Oncogene 2004; 23(36):6170-6174.
26. You Z, Saims D, Chen S et al. Wnt signaling promotes oncogenic transformation by inhibiting c-Myc-induced apoptosis. J Cell Biol 2002; 157(3):429-440.
27. Chen S, Guttridge DC, You Z et al. Wnt-1 signaling inhibits apoptosis by activating beta-catenin/T-cell factor-mediated transcription. J Cell Biol 2001; 152(1):87-96.
28. Hay BA, Huh JR, Guo M. The genetics of cell death: approaches, insights and opportunities in Drosophila. Nat Rev Genet 2004; 5(12):911-922.
29. Timmer JC, Salvesen GS. Caspase substrates. Cell Death Differ 2007; 14(1):66-72.
30. Hersey P, Zhuang L, Zhang XD. Current strategies in overcoming resistance of cancer cells to apoptosis melanoma as a model. Int Rev Cytol 2006; 251:131-158.
31. Riedl SJ, Shi Y. Molecular mechanisms of caspase regulation during apoptosis. Nat Rev Mol Cell Biol 2004; 5(11):897-907.
32. Willis SN, Adams JM. Life in the balance: how BH3-only proteins induce apoptosis. Curr Opin Cell Biol 2005; 17(6):617-625.

33. Cordero J, Jassim O, Bao S et al. A role for wingless in an early pupal cell death event that contributes to patterning the Drosophila eye. Mech Dev 2004; 121(12):1523-1530.
34. Tomlinson A. Patterning the peripheral retina of the fly: decoding a gradient. Dev Cell 2003; 5(5):799-809.
35. McCartney BM, Dierick HA, Kirkpatrick C et al. Drosophila APC2 is a cytoskeletally-associated protein that regulates wingless signaling in the embryonic epidermis. J Cell Biol 1999; 146(6):1303-1318.
36. Yu X, Waltzer L, Bienz M. A new Drosophila APC homologue associated with adhesive zones of epithelial cells. Nat Cell Biol 1999; 1(3):144-151.
37. Ahmed Y, Nouri A, Wieschaus E. Drosophila Apc1 and Apc2 regulate Wingless transduction throughout development. Development 2002; 129(7):1751-1762.
38. Brunner E, Brunner D, Fu W et al. The dominant mutation Glazed is a gain-of-function allele of wingless that, similar to loss of APC, interferes with normal eye development. Dev Biol 1999; 206(2):178-188.
39. Freeman M, Bienz M. EGF receptor/Rolled MAP kinase signalling protects cells against activated Armadillo in the Drosophila eye. EMBO Rep 2001; 2(2):157-162.
40. Blair NP, Trempe CL. Hypertrophy of the retinal pigment epithelium associated with Gardner's syndrome. Am J Ophthalmol 1980; 90(5):661-667.
41. Buettner H. Congenital hypertrophy of the retinal pigment epithelium. Am J Ophthalmol 1975; 79(2):177-189.
42. Traboulsi EI, Murphy SF, de la Cruz ZC et al. A clinicopathologic study of the eyes in familial adenomatous polyposis with extracolonic manifestations (Gardner's syndrome). Am J Ophthalmol 1990; 110(5):550-561.
43. Marcus DM, Rustgi AK, Defoe D et al. Retinal pigment epithelium abnormalities in mice with adenomatous polyposis coli gene disruption. Arch Ophthalmol 1997; 115(5):645-650.
44. Dorsky RI, Moon RT, Raible DW. Environmental signals and cell fate specification in premigratory neural crest. Bioessays 2000; 22(8):708-716.
45. Ikeya M, Lee SM, Johnson JE et al. Wnt signalling required for expansion of neural crest and CNS progenitors. Nature 1997; 389(6654):966-970.
46. Lee HY, Kleber M, Hari L et al. Instructive role of Wnt/beta-catenin in sensory fate specification in neural crest stem cells. Science 2004; 303(5660):1020-1023.
47. Garcia-Castro MI, Marcelle C, Bronner-Fraser M. Ectodermal Wnt function as a neural crest inducer. Science 2002; 297(5582):848-851.
48. Dorsky RI, Moon RT, Raible DW. Control of neural crest cell fate by the Wnt signalling pathway. Nature 1998; 396(6709):370-373.
49. Hasegawa S, Sato T, Akazawa H et al. Apoptosis in neural crest cells by functional loss of APC tumor suppressor gene. Proc Natl Acad Sci USA 2002; 99(1):297-302.
50. van de Wetering M, Sancho E, Verweij C et al. The beta-catenin/TCF-4 complex imposes a crypt progenitor phenotype on colorectal cancer cells. Cell 2002; 111(2):241-250.
51. Batlle E, Henderson JT, Beghtel H et al. Beta-catenin and TCF mediate cell positioning in the intestinal epithelium by controlling the expression of EphB/ephrinB. Cell 2002; 111(2):251-263.
52. de Lau W, Barker N, Clevers H. WNT signaling in the normal intestine and colorectal cancer. Front Biosci 2007; 12:471-491.
53. Wong MH, Rubinfeld B, Gordon JI. Effects of forced expression of an NH2-terminal truncated beta-Catenin on mouse intestinal epithelial homeostasis. J Cell Biol 1998; 141(3):765-777.
54. Sansom OJ, Reed KR, Hayes AJ et al. Loss of Apc in vivo immediately perturbs Wnt signaling, differentiation and migration. Genes Dev 2004; 18(12):1385-1390.
55. Wong MH, Huelsken J, Birchmeier W et al. Selection of multipotent stem cells during morphogenesis of small intestinal crypts of Lieberkuhn is perturbed by stimulation of Lef-1/beta-catenin signaling. J Biol Chem 2002; 277(18):15843-15850.
56. Kielman MF, Rindapaa M, Gaspar C et al. Apc modulates embryonic stem-cell differentiation by controlling the dosage of beta-catenin signaling. Nat Genet 2002; 32(4):594-605.
57. Gaspar C, Fodde R. APC dosage effects in tumorigenesis and stem cell differentiation. Int J Dev Biol 2004; 48(5-6):377-386.
58. Katoh M. Frequent up-regulation of WNT2 in primary gastric cancer and colorectal cancer. Int J Oncol 2001; 19(5):1003-1007.
59. Katoh M. Expression and regulation of WNT1 in human cancer: up-regulation of WNT1 by beta-estradiol in MCF-7 cells. Int J Oncol 2003; 22(1):209-212.
60. Wong SC, Lo SF, Lee KC et al. Expression of frizzled-related protein and Wnt-signalling molecules in invasive human breast tumours. J Pathol 2002; 196(2):145-153.
61. Suzuki H, Watkins DN, Jair KW et al. Epigenetic inactivation of SFRP genes allows constitutive WNT signaling in colorectal cancer. Nat Genet 2004; 36(4):417-422.

62. Smith K, Bui TD, Poulsom R et al. Up-regulation of macrophage wnt gene expression in adenoma-carcinoma progression of human colorectal cancer. Br J Cancer 1999; 81(3):496-502.
63. Dimitriadis A, Vincan E, Mohammed IM et al. Expression of Wnt genes in human colon cancers. Cancer Lett 2001; 166(2):185-191.
64. Holcombe RF, Marsh JL, Waterman ML et al. Expression of Wnt ligands and Frizzled receptors in colonic mucosa and in colon carcinoma. Mol Pathol 2002; 55(4):220-226.
65. Satoh S, Daigo Y, Furukawa Y et al. AXIN1 mutations in hepatocellular carcinomas and growth suppression in cancer cells by virus-mediated transfer of AXIN1. Nat Genet 2000; 24(3):245-250.
66. Chen T, Yang I, Irby R et al. Regulation of caspase expression and apoptosis by adenomatous polyposis coli. Cancer Res 2003; 63(15):4368-4374.
67. Morin PJ, Vogelstein B, Kinzler KW. Apoptosis and APC in colorectal tumorigenesis. Proc Natl Acad Sci USA 1996; 93(15):7950-7954.
68. Huang M, Wang Y, Sun D et al. Identification of genes regulated by Wnt/beta-catenin pathway and involved in apoptosis via microarray analysis. BMC Cancer 2006; 6:221.
69. Ioannidis V, Beermann F, Clevers H et al. The beta-catenin—TCF-1 pathway ensures CD4(+)CD8(+) thymocyte survival. Nat Immunol 2001; 2(8):691-697.
70. Dikovskaya D, Schiffman D, Newton IP et al. Loss of APC induces polyploidy as a result of a combination of defects in mitosis and apoptosis. J Cell Biol 2007; 176(2):183-195.
71 Qian J, Steigerwald K, Combs KA et al. Caspase cleavage of the APC tumour suppressor and release of an amnio-terminal domain is required for the transcription-independent function of APC in apoptosis. Oncogene 2007; 26:4872-4876.

Chapter 8

APC and Its Modifiers in Colon Cancer

Lawrence N. Kwong and William F. Dove*

Abstract

Colon cancer closely follows the paradigm of a single "gatekeeper gene." Mutations inactivating the *APC* (*adenomatous polyposis coli*) gene are found in ~80% of all human colon tumors and heterozygosity for such mutations produces an autosomal dominant colon cancer predisposition in humans and in murine models. However, this tight association between a single genotype and phenotype belies a complex association of genetic and epigenetic factors that together generate the broad phenotypic spectrum of both familial and sporadic colon cancers. In this Chapter, we give a general overview of the structure, function and outstanding issues concerning the role of Apc in human and experimental colon cancer. The availability of increasingly close models for human colon cancer in genetically tractable animal species enables the discovery and eventual molecular identification of genetic modifiers of the *Apc*-mutant phenotypes, connecting the central role of Apc in colon carcinogenesis to the myriad factors that ultimately determine the course of the disease.

Colorectal Cancer

Colorectal cancer is the second leading cause of cancer morbidity and mortality worldwide.[1] Almost half of the population will develop at least one benign adenomatous colonic polyp during life, with less than 3% of those cases going on to develop colorectal cancer. Because symptoms are rare until very late stages, most cases go undetected. Colon cancer manifests itself as polypoid growths that progress to malignancy; metastases to the lymph nodes, liver and lung are the primary cause of death in patients with advanced disease.

In the study of colon cancer, research is divided between sporadic and familial cases. Although hereditary colon cancer predispositions make up less than 5% of all colon cancer cases worldwide, the extensive pedigree information available in such cases has provided statistical power for isolating both the underlying causes and the genetic, environmental and dietary modifiers of the phenotypes. The relationship of sporadic to familial colon cancer is highlighted by the successful use of therapeutics such as nonsteroidal anti-inflammatory drugs (NSAIDs) to treat both diseases.[2] At present, a combination of chemotherapy, radiation treatment and surgery is used to treat colon cancer. The 5-year survival expectation for colon cancer patients ranges from 93% for early stages to 8% in fully advanced stages.[3]

In this chapter, we will introduce and review the genetics and function of the central gatekeeper gene in colon cancer: *Adenomatous Polyposis Coli* (*APC*/*Apc*).[a]

[a] *APC* and *Apc* are the designations for the human and murine genes, respectively; Apc is used herein for the function of the gene, regardless of species.

*Corresponding Author: William F. Dove, McArdle Laboratory for Cancer Research, University of Wisconsin, 1400 University Avenue, Madison, Wisconsin 53706, USA.
Email: dove@oncology.wisc.edu.

APC Proteins, edited by Inke S. Näthke and Brooke M. McCartney.

Biology of the Human Intestine

The small intestine is composed of interdigitated villi and crypts of Lieberkühn (for a more in-depth discussion, see Sansom, this volume). The villi serve an absorptive function in the processing of food.[4] The colon does not contain villi, but rather is composed of crypts, invaginated into a flat surface that is folded at various intervals called rugae. During human development, the adult intestine expands in part by a process of crypt fission, where entire crypts divide, producing daughter crypts.[5,6] This process "purifies" crypts in that the early polyclonal crypts[7] become monoclonal. Thus, each adult crypt lineage is limited to one somatic genotype. Crypt purification also occurs by stem cell succession, whereby a clone becomes dominant within the crypt. Analysis of methylation patterns in human crypts shows that stem cell succession continues over the life of an adult, as measured by random methylation changes that gradually become fixed in a crypt.[8] An estimated 4-16 adult stem cells reside as a clonal cohort in a niche near the bottom of each crypt. As cells reach the top of the villus in the small intestine or the collar of the crypt in the colon, they undergo apoptosis and are shed into the intestinal lumen. Cells of crypts thus turn over at a high rate (every 3-5 days[9]) owing to the continual flow of newly produced cells up the crypt/villus axis (see Potten and Morris[10] for a review of a classic body of work).

Intestinal epithelial stem cells can differentiate into a number of different cell types.[11,12] Within colonic crypts lie goblet cells that secrete mucus; at the base of small intestinal crypts lie Paneth cells that provide defense and that help to maintain the gut flora. Enterocytes perform an absorptive function for nutrients crossing the epithelium and comprise up to 80% of the small intestine. Finally, rare enteroendocrine cells, comprising ~1% of the intestine, secrete hormones such as serotonin. Below the epithelial layer lies the *lamina propria*, which comprises the stromal connective and endothelial tissue that lends support and circulation to the epithelial cells. The *muscularis mucosa* lies immediately below the epithelial layer and separates it from the submucosa, which is composed of connective tissue. Below that is the *muscularis externa*, the muscle layer along which peristalsis moves food through the intestinal tract. Finally, the serosal layer marks the outermost edge of the intestine and is attached to the mesentery.

Development of Human Intestinal Tumors

Intestinal tumors have been hypothesized to arise from the stem cells near the bottom of crypts, but other interpretations are possible, as discussed below. Accumulating evidence in various fields of cancer research supports the stem cell origin of tumors.[13] Such research began with the study of hematopoietic stem cells, for which the genetics and quantitative biology had been well-established for several decades.[14] It was noticed that the cells of hematopoietic malignancies exhibited similarities to multipotential hematopoietic precursors, particularly the ability to self-renew.[15] Eventually, it was discovered that only a certain subpopulation of hematopoietic cancer cells are capable of transferring cancer to immunocompromised NOD/SCID mice.[16] Recently, solid tumors have been investigated in a similar manner. For example, human breast cancers passaged serially through NOD/SCID mice show that a small number of cancer cells expressing a certain profile of surface markers are sufficient to initiate new tumors, whereas a large number of cancer cells with different profiles are not sufficient.[17] Such "cancer stem cell" profiles have been shown for other cancer types including myeloma, brain and prostate.[18-20] Indeed recent studies have identified CD133 as a marker enriched in a self-renewing, tumor-initiating subpopulation of cells from human colonic tumors.[21,22] Progress in the development of diagnostic cell markers will help to resolve the issue of whether the genetic event that initiates tumorigenesis necessarily occurs in stem cells proper or whether, alternatively, they can also occur in undifferentiated or dedifferentiated daughter cells.[23]

The issue of tumor progenitor cells has led to a debate about whether intestinal tumors form by a "bottom-up" process originating at the stem cell niche, or by a "top-down" process originating in cells in the inter-cryptal space at the top of the crypt/villus axis. Evidence for the "top-down" hypothesis comes from monocryptal human sporadic adenomas in which dysplasia is confined to the top half of the crypt, with normal-appearing cells more basally located in the crypt.[24] The implication is that the dysplasia must have started at the top and grown down towards, rather than emerging from the stem cell niche. However, it is possible that the dysplasia originated in the middle of the crypt and expanded upwards. Thus, both the "bottom-up" and "top-down" models could be explained by an upwards expansion of stem cell derivatives[25] from the middle of the crypt, or by the transformation of daughter stem cells to becoming tumor-competent. Clearly, the molecular identification of colon cancer stem cells is needed to determine the location of the cell of origin for particular intestinal tumors.

An early stage of colonic tumorigenesis is the benign adenoma that progresses to adenocarcinoma in situ—tumors that have developed high-grade dysplasia but are confined to the region above the submucosa. Progression to adenocarcinomas with invasion into or beyond the submucosa can be classified using different systems. The Dukes staging system (Dukes A, B, C, D, or E) is a measure of how far the invasive front of the cancer penetrates the intestinal wall.[26] In the AJCC/TNM system, numbers identifying T (tumor), N (metastasis to the nodes) and M (metastasis to distant sites) provide a comprehensive view of tumor progression.[27] For example, a $T_4N_1M_0$ cancer indicates an adenocarcinoma that has invaded through the wall of the intestine and spread to 1-3 regional lymph nodes, but not yet to distant sites. Finally, the histological classification of polyps can be villous, tubulovillous, tubular, hyperplastic, or serrated. The rare villous adenoma class is believed to have the greatest potential for malignancy.[28] Hyperplastic and serrated polyps have traditionally been viewed as benign; however, recent evidence points to a possible hyperplastic-serrated-adenocarcinoma progression sequence that involves somatic hyperactivation of the BRAF oncogene.[29] The combination of these classification systems allows for a standardization of terminology among physicians. However, not all tumors fall into only one class and even tumors in the same nominal class can behave differently between and within patients.

Discovery of *APC* Mutations in Human Colon Cancer

Familial adenomatous polyposis (FAP) was first described as Gardner's syndrome[30] and included extracolonic manifestations such as osteomas and congenital hypertrophy of the retinal pigment epithelium (CHRPE). Over time, it became clear that different classes of FAP existed with different symptoms, of which Gardner's syndrome was only one. For example, "classical" FAP manifests as one hundred or more polyps in the colon, usually developing by twelve years of age, whereas patients with fewer than a hundred polyps are classified as attenuated FAP (AFAP). Many extracolonic symptoms further subdivide FAP.[31]

Linkage studies and the FAP-associated interstitial 5q Herrera deletion narrowed the genetic region underlying FAP to the 5q21 subchromosomal region (Fig.1).[32,33] The *APC* gene was then linked to FAP concurrently by Kinzler et al,[34] Nishisho et al,[35] Joslyn et al[36] and Groden et al.[37] *APC* mutations were subsequently found in ~80% of sporadic colorectal tumors,[38] confirming that Apc acts as a central gatekeeper protein in colorectal tumorigenesis. *APC* mutations and hypermethylation have also been found in various other cancer types, including pancreatic and gastric cancers.[39,40]

Function of Apc

Soon after the discovery of the *APC* gene, the function of the gene product came under intense scrutiny. The crucial understanding of its function came concurrently from Su et al[41] and Rubinfeld et al[42] who identified the relationship between Apc and the regulation of β-catenin. We now know that the central lesions in both hereditary and sporadic colon tumors result in activation of the Wnt signaling pathway (see Kennell and Cadigan, this volume). In nearly all tumors, deactivating *APC* or *GSK3β* mutations or stabilizing *CTNNB1* (encoding β-catenin) mutations are present.[43] More specifically, the canonical tumor suppressor function of Apc is to form a "destruction complex" with Axin/Axin2 and GSK-3β that promotes the ubiquitination and subsequent proteasomal degradation of the oncogene β-catenin in the absence of Wnt signaling. Loss of Apc function results in an accumulation of β-catenin, which translocates to the nucleus and engages the Tcf/Lef transcription factor complex to activate transcription of a large number of target genes including cyclinD1, c-myc and CRD-BP.[44] The tumorigenic consequences of unregulated β-catenin activity may be related to both the direct stimulation of cellular growth and proliferation and to the disruption of differentiation programs.

In addition to its role in the Wnt signaling pathway, Apc also functions to promote microtubule stability in a number of cellular contexts. The impact of the disruption of this function on tumorigenesis is not well understood (see Caldwell and Kaplan, Morrison and Bahmanyar et al, this volume). However, it is worth noting that two groups have reported that stabilized β-catenin, expressed either from a conditionally activatable allele exposed to Cre or from a transgene, is sufficient to induce intestinal polyposis in mice,[45,46] suggesting that loss of the microtubule-binding functions of Apc is not absolutely required for early tumor formation. Furthermore, as discussed below, mice homozygous for the 1638T *Apc* allele lacking the microtubule- and EB1-binding domains of Apc, but not the β-catenin binding domains, do not develop tumors. Despite these findings, an attractive speculation is that the disruption of microtubule functions contributes to tumor progression rather than to tumor initiation. Investigation of this idea awaits analysis of the progression stages of colonic neoplasia and the construction of mouse lines in which only the C-terminus of *Apc* can be conditionally deleted.

Structure of *APC*

The human *APC* gene spans 58 kb, with a 15-exon coding region of 8529 bp encoding a 2843 amino acid (aa), 310 kD protein. Several exons exist 5′ of exon 1: 0.1, 0.2, 0.3,[47] BS[48] and possibly more. The extent to which these isoforms play a role, if any, in colon cancer is unknown; many appear to be neuron-specific.[49]

The canonical Apc transcript initiates at exon 1 and produces a protein with eight known functional sub-domains (Fig. 2). The majority of truncating mutations with severe phenotypes remove most of the β-catenin-binding "20 amino acid" (20aa) repeats (1256-2031aa).[50] Interestingly, more C-terminal truncations that remove only the Axin-binding SAMP repeats (1568-2053aa),[51] microtubule binding repeats (2220-2597aa),[52] EB1-binding domain (2670-2843aa) and/or PDZ domain (the C-terminal 73aa that mediates anchoring to the cytoskeleton)[53] generally have an attenuated phenotype. N-terminal truncations that apparently affect only the homodimerization domain (6-57aa), owing to bypass through the use of an internal translation restart site, likewise generally give attenuated phenotypes (see Fig. 1).[54] Mutations that truncate within the armadillo repeats (453-767aa)—which bind several proteins including Asef and KAP3, both involved with different aspects of cytoskeletal function[55,56]—or within the β-catenin-binding "15 amino acid" (15aa) repeats (1021-1187aa) tend to be somewhat milder than the 20aa repeat truncations. An interesting molecular correlation in tumors was observed that may explain these findings: germline *APC* mutations in the mutation cluster region (MCR) spanning most of the 20aa repeats generally exhibit acquired loss of the wildtype allele, while *APC* mutations outside of this region generally exhibit acquired truncating mutations in the wildtype allele.[57] Several hypotheses have been put forth: the "just-right"[58] and "loose fit"[59] hypotheses, each of which proposes that an optimal number of 15aa repeats must remain after biallelic *APC* inactivation to produce a severe FAP phenotype. These hypotheses remain to be rigorously tested.

Genotype-Phenotype Correlation in FAP

One difficulty in understanding the genotype-phenotype correlation is the current lack of a comprehensive public database of FAP patients. For research on mouse models, this lack of data makes it difficult to contextualize observations in terms of the human disease. So far, a literature search has found only one large-scale attempt to compile such information, although it presents only the results of the analysis and does not make the raw data available.[60] Compounding this difficulty is that most reports on human cases do not count the multiplicity of tumors, but rather give only an estimate. Further difficulties come from differences in phenotype that may relate to whether the patient has received surgery or chemotherapeutics and to the age of diagnosis. To address this gap temporarily, we have compiled data on 441 cases from 37 reports (see http://mcardle.oncology.

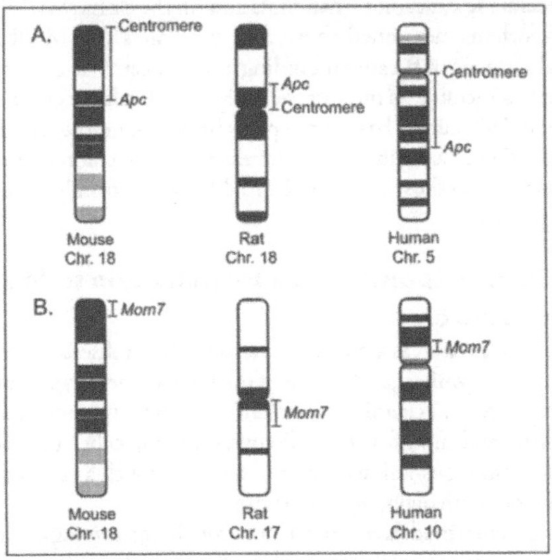

Figure 1. Organization of the mouse, rat and human chromosomes bearing the *APC/Apc* and *Mom7* orthologs. A) The *Apc* locus of the mouse lies on a telocentric chromosome, in contrast to its orthologs in the rat and human, each of which lies on a metacentric chromosome. A metacentric character enables a facile discrimination between whole chromosome loss versus somatic recombination. B) The *APC/Apc* locus of the mouse is linked to the *Mom7* locus on Chr 18, while the orthologs of these two loci are not linked in the rat and human karyotypes.

wisc.edu/dove/Data/FAP.htm). We suggest that a curated public database be generated under the aegis of a society for gastroenterology, for easy access to vetted information of this sort.

These data lead to a conclusion different from that of Crabtree et al,[60] who claim that "mutations between codons 1020 and 1169 hav[e] the mildest disease" and that the most N-terminal truncations (i.e., prior to codon 248) do not lead to an attenuated phenotype. Instead, it seems that N-terminal truncations produce the mildest disease, although mutations between codons 1020

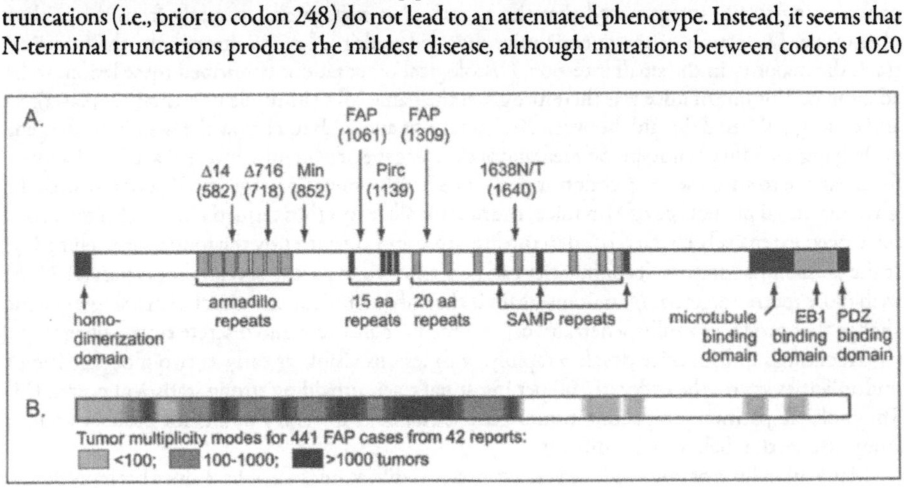

Figure 2. The structure of the Apc protein. A) Arrows on top indicate orthologous locations of mouse model mutations and the two most common FAP mutation sites. B) The genotype-phenotype correlation of sites of protein truncation to disease severity. The data used to generate this bar can be found at http://www.mcardle.wisc.edu/dove/Data/Apc.htm. aa, amino acid. Adapted from Amos-Landgraf and Kwong et al, Proc Natl Acad Sci USA 2007; 104:4036-4041.

and 1169 tend to generate fewer tumors than mutations in the classic MCR (codons 1250-1450; cf. the "loose fit" hypothesis mentioned above, which predicts that MCR mutations leave behind a more optimal number of β-catenin-binding 15aa repeats). These discrepancies could be explained by geographic ancestry, as most of the patients of Crabtree et al's come only from the UK, whereas our compiled data are based on reports from around the world. In this regard, it is interesting to note the significant differences in presentation of colonic cancer in patients from the Middle East compared to those from the United States,[61] possibly indicative of segregating modifier alleles (see below).

Biology of the Murine Intestine: An Introduction to Murine Models of Colon Cancer

The mouse has long been used as a model for various human diseases, due to its experimental tractability and frequently significant reflection of the human phenotype. For colon cancer, mice readily form polyps after certain chemical treatments or genetic modifications and have been an invaluable tool for drug and modifier locus discovery, among other benefits. In the following sections, we introduce numerous well-used mouse models, as well as a novel rat model. We also discuss other animal models involving *Apc* inactivation.

One caveat in using animal models is the deviation from human biology. The murine intestine—both mouse and rat—generally resembles that of the human in both development and structure, particularly in the formation of crypts and villi in the small intestine and in the crypt architecture of the colon. However, a few major differences exist: (i) the murine colon and small intestine are intermingled within the peritoneum, rather than separated, (ii) the rugae of the proximal murine colon have a diagonal rather than perpendicular pattern and (iii) the murine cecum is proportionately much larger. The extent to which these differences affect tumorigenesis is unknown, but must be taken into consideration when extrapolating from model animals to humans.

Mouse Models of Intestinal Cancer

The first hereditary mouse model of colon cancer was described in 1990. Efficient ENU mutagenesis of the germline of C57BL/6J (B6) mice and subsequent outcrossing to AKR/J mice identified a phenodeviant with both a circling behavior and anemia.[62] After continually backcrossing to B6, it was noted that the anemia trait segregated separately from the circling phenotype. Dissection of the anemic mice revealed multiple lesions throughout the intestinal tract, the majority in the small intestine. Histological preparations confirmed these lesions to be adenomas. This line of mice was therefore given the name Min (Multiple intestinal neoplasia). Su and colleagues[63] used the link between *Apc* mutations and FAP to narrow the search for the gene underlying the Min phenotype. Sequencing of the *Apc* gene of Min mice revealed a single change—from leucine to an amber stop codon at position 850. This mutation segregated perfectly with the small intestinal phenotype of Min mice; the mutant allele was thus termed *Apc*^{Min}. Min mice have since been extensively characterized in the literature and are currently the fourth best-selling line at the Jackson Laboratory. Its popularity can be attributed in part to several properties: (i) Along with more recent targeted *Apc* mutants, Min is the only mouse cancer model with a single genetic change that produces a fully penetrant, organ-specific, consistent and discrete tumor phenotype. (ii) Adenomas in Min mice develop rapidly, with lesions visible as early as two months. Tumor multiplicities are on the order of 100 per intestinal tract, providing strong statistical power. (iii) The multiple pathways impacting tumorigenesis enable many entry points for basic or applied study (see section below on modifiers).

Many other lines of mice with targeted genetic modifications of *Apc* have since been produced. Table 1 provides a summary of mice generated with these disruptions. When heterozygous, the Δ474, Δ14, Δ716, lacZ and Δ1309 models all give phenotypes similar to that of Min.[64-68] In contrast, heterozygosity for the 1638N allele results in 0-2 tumors (none in the colon)[69] while the 1638T model is tumor-free and, unlike any other truncating allele, is homozygous viable.[70] Each of these two alleles truncates the protein at amino acid 1638; however, 1638N has only approximately 2%

the transcript expression level of wild type Apc while 1638T has the full expression level. The latter observation implies that the C-terminus of Apc containing the direct microtubule and PDZ binding domains is nonessential, either for normal embryonic development or for preventing tumor initiation. However, it is important to note that the 1638T allele is not completely wildtype, since animals doubly heterozygous for 1638T and Min are embryonic lethal (as discussed by Sansom, this volume). Nonetheless, the two observations suggest that it is the reduction in Apc protein, not the codon 1638 truncation itself, which results in the 1638N tumor phenotypes. That a reduction in functional Apc protein levels leads to tumor initiation was confirmed by Li and colleagues,[71] who inserted a neomycin cassette in either orientation (reverse, neoR, or forward, neoF, see Table 1) into the 13th intron of *Apc* to generate full-length hypomorphic alleles. These heterozygous mice developed fewer than two adenomas per mouse, with Apc protein levels and activity (as measured by β-catenin transcriptional activity) inversely correlating with tumor multiplicity. However, it is unclear whether the neomycin/hygromycin cassette in these insertion alleles of Fodde et al and Li et al exerts a regional position effect on a neighboring gene(s) that may also contribute to the phenotype.[72] In this context, a clear demonstration of modification of the Min phenotype by a cis-linked recessive lethal factor has been provided in the analysis of the modifier locus *Mom2*.[73]

Recent advances in molecular cloning have enabled the construction of three independent conditional alleles of *Apc* in which specific exons are flanked by *loxP* sites (see Table 1): one allele that removes exon 11 upon the administration of Cre recombinase, resulting in truncation at codon 468[74] and two alleles that remove exon 14, resulting in truncation at codon 580.[65,75] The homozygous ablation of *Apc* in various organs has broadened the understanding of the known functions of *Apc* in maintaining homoeostasis in the liver, kidney, thymus and intestine.[74,76-79] Indeed, carcinomas are induced in the liver and kidney upon tissue-specific deletion of *Apc*. The ability to temporally control *Apc* loss, combined with a titration of Cre, opens up novel avenues for understanding the sufficiency of Apc loss for tumorigenesis. The recent finding that somatic c-*Myc* deletion abrogates the phenotype of concomitant *Apc* loss in the intestine confirms the power of such conditional alleles for pathway analysis.[80]

Finally, chemical carcinogens such as AOM[81] and ENU[82] have been shown to induce intestinal cancer in wild type mice and have been used as models of colon cancer.[83]

Biology of Mouse Intestinal Tumors

Tumors in the small intestine of the Min mouse are composed of dysplastic crypts surrounded and supported by hyperplastic villi and crypts, displaying a characteristic "rose" shape. By contrast, colonic tumors are peduncular, forming a spherical mass of dysplastic cells supported by a stromal stalk.[84] Tumors have a higher mitotic index than adjacent normal tissue[85] and crypt fission indices of Min intestines are also higher than wild type.[5] In contrast to the top-down/bottom-up controversy in human tumorigenesis,[86,24] reviewed by Leedham and Wright,[87] there is little controversy over the directionality of tumor development in the Min or Δ716 mouse models: tumors begin as an outpocketing in the crypt and the dysplastic cell population expands in both directions along the crypt/villus axis.[84]

Rat Models of Intestinal Cancer

Wild type rats develop colon cancer at a very low incidence (<0.1%)[88] with the exception of the Wistar-Furth/Osaka line that spontaneously develops adenocarcinomas at a rate of 30-40%.[89] However, the genetic factors underlying this predisposition are unknown and no recent studies have been reported. The majority of current rat models of colon cancer rely on the induction of tumors via treatment with the carcinogens AOM, DMH, or PhIP.[90] The advantages of carcinogen-treated rat models are that tumors often progress to adenocarcinomas and that tumors have not been reported in the small intestine; the disadvantages are low polyp multiplicities (<2 in F344), long tumor latencies (>10 months) and laborious carcinogen administration regimens with the potential for inconsistent dosage. Carcinogen treatments have been required in the past, owing to the lack of rat embryonic stem cells required for generating genetically engineered rats. However, the ability

Table 1.* Apc *mutant mouse lines

Allele	Truncation Codon	Conditional?	Genetic Background	Intestinal Tumor #	% of wt Protein per Allele	Initial Reference of Phenotype
Δ468	468 (armadillo repeats)	Yes	N/Av	N/Av	N/Av	Gounari et al, 2005[74]
Δ474	474 (armadillo repeats)	No	B6	>100	100	Sasai et al, 2000[64]
Δ14	580 (armadillo repeats)	Yes	B6	>100	100	Colnot et al, 2004[65]
580S	580 (armadillo repeats)	Yes	Mixed	N/Av	N/Av	Shibata et al, 1997[75]
Δ716	716 (armadillo repeats)	No	B6	>100	0*	Oshima et al, 1995[149]
lacZ	716 (armadillo repeats)	No	Mixed	>100	100	Ishikawa et al, 2006[150]
Min	850 (armadillo repeats)	No	B6	>100	100	Moser et al, 1990[62]
Δ1309	1309 (15aa repeats)	No	B6	40	100	Niho et al, 2003[68]
1638N	1638 (SAMP repeats)	No	B6	1	2	Fodde et al, 1994[69]
1638T	1638 (SAMP repeats)	No	B6	0	100	Smits et al, 1999[70]
Ex13 NeoR	Full-length	No	B6	1	20	Li et al, 2005[71]
Ex13 NeoF	Full-length	No	B6	0.3	10	Li et al, 2005[71]

*This is suggested, but not proven.[71] N/Av = not available.

to generate target-selected mutations, including nonsense alleles, has recently been implemented by several laboratories.[91,92] This capacity has been drawn upon to generate a rat strain carrying a nonsense allele in codon 1137 of *Apc*. F344 rats heterozygous for this allele develop multiple intestinal neoplasms by three months of age, predominantly in the colon and survive in the range of one year.[93] The important colonic predisposition of tumorigenesis in this strain has led to its designation as Pirc: polyposis in the rat colon.

The size of the laboratory rat confers certain advantages to the Pirc model; for one, classical endoscopy can be used to monitor and biopsy colonic tumors.[93] In addition, microCT and micro-PET imaging can strengthen the annotation of each of the tumors, whose sizes—often exceeding 1 cm in diameter—greatly facilitate visualization and biopsy sampling. It can significantly enhance the molecular and morphological analysis of tumor progression to annotate individual neoplasms while keeping the animal alive. While these methods are also feasible in mouse models of colon cancer, the colonic predisposition, size and longevity of the tumor-bearing Pirc rat can provide significant advantages in developing these experimental avenues. Thus, the rat's promising utility for genetics combined with its size and feasibility for longitudinal studies of therapeutic regimes poises the Pirc kindred as a model for colon cancer that is complementary to the genetically powerful Min mouse model.

Coincidentally, the rat and mouse *Apc* loci each lie on Chromosome 18 of their respective genomes. The synteny over Chromosome (Chr) 18 is remarkably conserved between the mouse and the rat. The only difference in synteny is the most proximal 10 Mb of the mouse chromosome, the homologous region of which is located on rat Chr 17. However, a more important difference between these two versions of Chr 18 is the placement of the centromere. *Apc* lies ~30 Mb distal of the acrocentric mouse centromere but ~11 Mb proximal of the metacentric rat centromere (Fig. 1). By contrast, in the metacentric human Chr 5, *Apc* is ~65 Mb distal of the centromere.

Apc Mutations in Other Organisms

To date, *Apc* mutants have been isolated in three other experimental organisms. The $Apc^{MCR/+}$ zebrafish (*Danio rerio*) develops intestinal, hepatic and pancreatic neoplasms, demonstrating the conservation of organ-specific gene functions between vertebrate phyla.[94] *Drosophila melanogaster* lines heterozygous for mutations in either of the two *Apc* homologs, *dApc1* or *dApc2*, develop with a completely normal phenotype despite the evolutionary conservation of Wnt signaling function.[95] It is interesting to note in this context that *dApc1* can complement the function of human *Apc* in suppressing β-catenin-mediated transcription in colon cancer cell lines.[96] Finally, RNAi-induced reduction of *Caenorhabditis elegans Apr-1*, a gene homologous to the N-terminal half of human *APC*, results in aberrations in blastomere development and endoderm specification.[97] Recent studies have linked Wnt signaling and the regulation of WRM-1, a nematode homolog of β-catenin, to Apr-1 function during critical asymmetric cell divisions in development.[98]

Mechanisms of Loss of Heterozygosity at the *Apc* Locus

Biallelic loss of *Apc* function appears to be required for tumorigenesis, but it remains open whether a heterozygous phenotype (also see below) is a necessary preliminary step to the complete loss of Apc function in tumors. In principle, loss of function of the wild type allele from the heterozygote can occur through any of several mechanisms, including: somatic recombination, nondisjunction with or without reduplication, coding or regulatory mutations, epigenetic silencing, or partial or full gene deletion. Early studies in Min mice demonstrated whole-chromosome loss of heterozygosity (LOH),[99] narrowing the possibilities to somatic recombination or nondisjunction. However, the acrocentric nature of mouse chromosomes makes it difficult to distinguish between somatic recombination, which results in the homozygosis of all alleles distal to the recombination site and mitotic nondisjunction, which results in the loss of an entire homolog. Unless the centromere can be marked, each of these processes gives identical results for acrocentric, but not for metacentric chromosomes. Subsequent studies in Min mice harboring an abnormal Robertsonian metacentric Chromosome 18,[100] in Pirc rats with a naturally metacentric Chromosome 18 (Fig. 1),[93] and in FAP patients with

Apc truncations past codon 1286,[101] are consistent with somatic recombination; the majority of these intestinal tumors exhibit LOH limited to a single chromosome arm. Further, the genomes of the early mouse tumors appear to be stable, as assessed by FISH and karyotypic analysis.[102] Somatic recombination has also been shown to be involved in LOH of other tumor suppressors in humans, such as the retinoblastoma gene *Rb1*.[103,104]

By contrast, analysis of sporadic rather than familial human colon tumors suggests that the loss event may occur via a karyotypically unstable pathway. For example, Thiagalingam and colleagues[105] demonstrated that the observed single *p*-arm loss seen in 36% of tumors involved complex translocations rather than conservative somatic recombination. However, it is unclear whether the translocations were the cause of LOH, or instead were acquired during tumor progression. A study by Shih and colleagues,[106] showed allelic imbalance across the genome by digital SNP analysis; however, this finding will require confirmation using more current technology such as Pyrosequencing.[107] Another study has shown that 1638N tumors exhibit significant genomic copy number changes by comparative genomic hybridization;[108] this highlights differences between the 1638N and the genomically stable Min models since the 1638N phenotype may be influenced by regional position effects from the neomycin cassette.[72] In these investigations, another open issue is whether the earliest stage in tumorigenesis is being analyzed. Thus, the debate over the role of genomic instability in colorectal tumorigenesis remains divided into two hypotheses: that instability is a prerequisite for initiation and will be observed at the "birth" of the neoplasm, or that it is acquired during dysplastic growth along the neoplastic pathway and necessary only for progression.

Mathematical models have been invoked to support each hypothesis. Nowak and colleagues[109] showed theoretically that chromosomal instability (CIN) can drive the majority of sporadic LOH events: a hypothesized efficient statistical "tunneling" effect of CIN could drive cells towards an equilibrated LOH population. By contrast, Komarova and Wodarz[110] suggested that CIN would not be efficient, owing to the lag time required for the initial genomic hit to create CIN. Furthermore, Tomlinson and colleagues[111] used an evolutionary approach to stem cell statistics to show that any instability associated with colonic tumors could be explained by a selective, exponential accumulation of aberrations, rather than by a pre-existing state of instability. Such mathematical models may prove to be valuable frameworks for the design of new quantitative experimental tests.

Are Some Apc Truncation Peptides Dominant Negative?

Several lines of evidence suggest that certain truncated Apc proteins might act in a dominant negative manner, either by homodimerizing to wild type Apc or by competing for binding to β-catenin. For example, transfection of constructs encoding the N-terminal 750aa, 1309aa, 1450aa, or 1807aa of human Apc into colorectal cancer cell lines induced chromosome segregation dysfunctions, even in diploid cell lines.[112,113] Another example is that endogenous N-terminal Apc fragments bind to exogenous C-terminal fragments, altering the former's ability to bind to its partner Kap3.[114] Thus, truncated Apc proteins could dominantly interfere with the function of the remaining allele's product. Less direct lines of evidence come from analysis of normal tissue in Min mice. For example, differences have been observed between the intestines of Min and wild type mice in enterocyte migration,[115] E-cadherin localization[116] and Egfr expression.[117] It is not yet resolved whether these effects are autonomous to the heterozygous normal tissue, or are caused by a systemic effect of the tumors carried in the Min mouse.

By contrast, a line of mice transgenic for a Δ716 or Δ1287 fragment of the *Apc* gene failed to develop intestinal tumors.[118] Here, it is unclear whether the transgene expression levels reached a tumorigenic threshold, especially in the presence of two copies of the wild type allele. The question of whether Min is dominant negative has important implications for the study of LOH. If normal heterozygous tissue from Min animals has a phenotype that predisposes to tumorigenesis, then the familial case may differ from the sporadic case, where normal tissue is homozygous wild type for Apc. A full understanding of Apc action must also account for the full-blown polyposis phenotype of locus-wide deletions including the classical Herrera deletion by which the *APC* locus was first mapped.[33,119,120] It is also worth noting that similar C-terminal truncations of APC2 in Drosophila do not exhibit dominant negative effects on Wnt signaling or viability, but in some cases do have dominant effects on cytoskeletal organization in the embryo.[95] Thus, the question of predisposing haploinsufficiency or dominant negativity requires resolution.

Modifiers of Murine Intestinal Cancer

Many different pathways have an impact on the initiation and/or progression of intestinal adenomas: karyotypic stability, DNA mutation rates, stem cell turnover, cellular growth and proliferation, cellular differentiation, environmental factors, diet, exercise, therapeutic drugs and others. In this chapter we address only genetic modifying factors (for a review of diet and therapeutic drugs, see ref. 93).[90] In experimental genetics, a modifying locus has no phenotypic consequence in the absence of mutation at the primary locus of interest, in this case *Apc*. In epidemiology, however, the factors controlled by modifying loci may be found to have an impact, since the functional state of the primary locus may vary covertly or overtly in the population being studied.

The phenotypic variation of Min among different inbred strains highlights the importance of modifier alleles. Historically, B6-Min mice develop approximately 100 tumors in the intestinal tract. Other inbred backgrounds on which the Apc^{Min} allele has been introgressed show a broad spectrum of tumor multiplicities (Table 2). For example, BTBR is a strongly enhancing background, with mice becoming moribund by 60 days of age due to the presence of more than 600 tumors.[121] At the other extreme lie AKR mice, which develop only one to four tumors per animal and can survive for up to a year of age.[122] C3H and 129S6 have milder suppressive phenotypes compared to AKR. General strain effects have led the way for the identification of polymorphic modifier loci by quantitative trait locus analysis of the phenotypes of *Min* carriers in outcrossed progeny.[123]

Perhaps the most well-known modifier is *Mom1* (Modifier of Min 1). A quantitative trait locus (QTL) analysis using SSLP markers in crosses involving 4 inbred strains found a QTL on chromosome 4 that was shared among all mapping crosses.[123] It was apparent that at least two alleles of *Mom1* existed: a resistance allele found in AKR/J, MA/MyJ and CAST/EiJ and a sensitivity

Table 2. The genetic background dependence of the Min phenotype

Strain	Mom1	Age (Days)	Small Intestine	Colon	N	Reference
129S6	S/S	103-163	45	1	23	L.N. Kwong, unpublished
BTBR/Pas	S/S	54-82	625	12	74	Kwong et al, 2007[121]
C3H/HeJ	S/S	100-120	16	0.4	89	Koratkar et al, 2004[151]
C57BL/6J	S/S	90-120	128	3	48	Kwong et al, 2007[121]
AKR/J	R/R	146-336	4	0	42	Kwong et al, 2007[121]
129 × B6 F1	S/S	92-164	82	0.2	35	L.N. Kwong, unpublished
AKR × B6 F1	R/S	104-143	25	0.1	15	Kwong et al, 2007[121]
BTBR × B6 F1*	S/S	80-93	117	1.6	16	A. Shedlovsky, unpublished
BTBR × B6 F1**	S/S	84-89	215	1.4	19	A. Shedlovsky, unpublished
C3H × B6 F1	R/S	130-150	8	0	10	Koratkar et al, 2004[151]
CAST × B6 F1	R/S	100-120	3	0	14	Koratkar et al, 2002[152]
CAST × B6 F1	R/S	185-215	7	0	11	Koratkar et al, 2002[152]

*Min from B6 parent. **Min from BTBR parent.

allele in C57B/6J (B6). *Mom1* is semidominant where each copy affects tumor number by a factor of about 2. MacPhee and colleagues[124] suggested that the *Pla2g2a* gene (encoding secretory phospholipase 2A) might explain the *Mom1* effect. This hypothesis was confirmed in a line of B6 Min mice transgenic for a cosmid containing the resistance allele *Pla2g2a*,[125] which showed reduced polyp number. Subsequent higher resolution genetic analysis showed that the *Mom1* locus consists of both *Pla2g2a* and at least one other distal factor.[126] The effect of *Mom1* explains a significant proportion of the variance in tumor multiplicity seen in crosses between B6-Min and AKR or C3H mice (Table 2). Interestingly, the *Pla2g2a* gene seems to act in a cell non-autonomous fashion: it is expressed from postmitotic Paneth and goblet cells within the micro-environment, affecting the net growth rate of adjacent tumors.[85] (Evidence has been reported that the secretory phopholipase A2 can instead stimulate colonic tumor growth when expressed autonomously within the tumor lineage.[127]) This apparent non-autonomous action of *Pla2g2a* illustrates the necessity of investigations in the whole animal, as such effects would be lost in cell culture or non-orthotopic xenograft models.[128] The exact mechanism by which *Pla2g2a* exerts its effects on colon tumorigenesis remains unresolved,[129] highlighting the challenges of cancer modifier genetics. Furthermore, its relevance to the human disease is unresolved. Three studies have failed to find significant cancer-associated germline or somatic variation in the human *PLA2G2A* gene.[130-132] One sporadic colon cancer patient has been reported with a constitutional frameshift mutation in this gene.[133] Finally, a correlation has been reported between PLA2G2A expression and gastric adenocarcinoma patient survival.[134] Overall, the identification of *Mom1* has had a long-lasting impact on modifier genetics, as it was an important proof of principle that such studies could identify at the molecular level genetic determinants modifying a cancer phenotype.

By utilizing similar mapping methods, additional polymorphic Modifiers of Min have been discovered: *Mom2*, *Mom3* and *Mom7*, each of which resides on Chromosome 18. *Mom2* arose spontaneously in a stock of $Apc^{Min/+}$ mice on the C57BL/6J background and mapped distal to the *Apc* locus.[135] Congenic line, expression and sequencing analyses pinpointed a recessive embryonic lethal 4 bp duplication in the ATP synthase *Atp5a1* gene.[73] When in cis with the mutant *Min* allele, this mutant *Mom2* allele confers an ~12-fold resistance to tumor multiplicity, but has no effect when in trans. Along with a decreased LOH incidence, these results indicated that somatic recombination proximal to both the *Apc* and *Atp5a1* loci would generate homozygous *Atp5a1* segregants that would be cell- and therefore tumor-lethal.

The *Mom3* locus was discovered in a line of Min mice that had become strain-contaminated,[136] resulting in an increase in tumor multiplicity compared to control B6-Min mice. It mapped to within the first 25cM of chromosome 18, proximal to *Apc*. However, the lack of additional polymorphic markers, along with the unknown contaminating strain background, prevented further positional refinement. In a separate study, the *Mom7* locus mapped to a similar region as *Mom3*, but came from defined crosses of the B6.$Apc^{Min/+}$ line to the AKR, BTBR and A/J strains.[121] Congenic line and in silico mapping analyses reduced the *Mom7* interval to the first 4.4 Mb of chromosome 18, including the complex sequence of the centromere. Unlike *Mom2*, *Mom7* is homozygous viable for all alleles and the B6 allele shows a dominant resistance phenotype in both the trans and cis configurations. Whether *Mom7* and *Mom3* represent the same underlying modifier must be resolved by complementation testing. Interestingly, the Rb(7.18)9Lub Robertsonian translocation (Rb9), also at pericentromeric Chromosome 18, lowers tumor multiplicity in $Apc^{Min/+}$ mice.[100] FISH analysis showed that the Chromosome 18 homologs were mispaired in the nucleolar organizing region, leading to the hypothesis that the opportunity for somatic recombination at *Apc* is decreased by this centric fusion. Although *Mom7* and Rb9 map to the same location, it is important to note that Rb9 involves a gross physical chromosome abnormality, while *Mom7* involves a normal chromosome; furthermore they have qualitatively different effects, with *Mom7* resistance fully dominant and Rb9 semidominant, making it unlikely that they represent the same modifier. Furthermore, none of these modifiers shows the "overdominant effect" predicted for sequence heterozygosity, which would suppress somatic recombination in heterozygotes but not in homozygotes.[137] Thus, the *Mom7* and *Mom3* are modifiers distinct from Rb9.

As illustrated by the growing set of modifiers of the Min phenotype, it is clear from Table 3 that strategies for cancer prevention and therapy have many points of entry, providing both a wealth of candidate therapeutic targets and the challenge of converting any of them into potential human therapies. However, the benefit of such modifier studies extends beyond clinical relevance; each dataset informs both the functions of the modifier and of *Apc*. In turn, each modifier has a role in processes other than tumorigenesis. For example, the increases in both karyotypic instability and tumor multiplicity in $BubR1^{+/-}$;$Apc^{Min/+}$ mice provide insight into the normal checkpoint functions of both BubR1 and *Apc*.[138] Another interesting example is that deletion of *H19* induces the biallelic expression of *Igf2*, increasing Min tumor multiplicities.[139] This genetic model of loss-of-imprinting (LOI) highlights the functional importance of genomic imprinting. In human sporadic colorectal cancer patients, LOI at *Igf2* is often elevated in peripheral blood lymphocytes compared to healthy controls,[140] implying that LOI can precede the loss of Apc function and become a risk factor for otherwise normal individuals.

Probing deeper into the modifiers organized in Table 3, several interesting patterns are noted. First, mutations in either of the mitotic stability genes *BubR1*[138] or *Cdx2*[141] generate a complex modifying phenotype, whereby the multiplicities of tumors of the small intestine decrease, while multiplicities of colonic tumors increase. This striking disparity between the effects of the same mutation in two different regions of the gut suggests that the small intestine and colon have different abilities to respond to CIN. Perhaps the small intestine expresses a senescence and/or apoptosis response that efficiently blocks CIN-induced tumor formation. By contrast, the hyper-recombination phenotypes of *Blm*[142,143] or *Reql*[144] mutations affect the entire intestinal tract.

The contrast between the regionally diverse response to mitotic instability and the uniform response to hyperrecombinational instability suggests that different responses to different types of instability exist in different regions of the intestinal tract. In the same vein, the Mbd2 and Mbd4 methyl-binding proteins have opposite effects on intestinal tumor multiplicity,[145,146] indicating that the epigenetic machinery has both positive and negative indirect regulators of methylation-associated DNA mutation and/or silencing. Indeed, the potency of mutations in mismatch repair genes to generate tumors in the ascending colon illustrates both the centrality of sequence stability to tumor suppression and the regionality of these effects. Next, mutations in the Ephrin family of genes[147] demonstrate that differentiation is key to tumorigenesis, mirroring the dysregulation of ephrin receptors in mice conditionally inactivated for *Apc*.[78] Finally, many "classic" regulators of numerous tumor pathways—including p53, p27, p21, c-Jun and cyclin D1—modify the Min phenotype, raising the possibility that therapies directed towards other classes of cancer could also have an effect on colonic tumors.

Conclusion

The complexity of both morphological and molecular pathways in colon cancer presents a challenge to clinical therapies, which are already multifaceted. For example, the FOLFOX regimen combines fluorouracil, leucovorin and oxaliplatin, which can be used in addition to standard surgery and radiation treatments. Despite the complexity, the many different animal models now available—mouse, rat, zebrafish and invertebrates—expand our ability to identify and validate different therapeutic targets. Indeed, the convenience of these animal models simplifies many aspects of colon cancer research that would otherwise be difficult to control from a highly heterogeneous human population. The effectiveness of such models emerged from the discovery of Apc as the central molecule negatively regulating colon cancer. This discovery, a result of Herculean efforts by several centers of human genetics[33,34,37,148] allowed for both the identification of the molecular basis of the Min phenotype and the characterization and construction of single-gene mutants with profound cancer phenotypes. Overall, the study of colon cancer radiates out from our understanding of the mechanisms of action of the Apc protein, a central node regulating multiple cancer pathways.

Table 3. Molecular genetic modifiers of Apc *knockout mouse models*

Modifier Affects	Modifier Gene(s)	Modifier Allele(s)	Allele Property	Apc Model	Effect of Mutant Allele on Intestinal Tumor Multiplicity	Factor of Effect	Reference
Karyotypic stability	BubR1	$Bub1b^{Gt(neo-btk)1Dai}$	Knockout (het)	Min	Decrease/Increase[a]	2/10[a]	Rao et al, 2005[138]
	Cdx2	$Cdx2^{tm1Mmt}$	Knockout (het)	Δ716	Decrease/Increase[a]	9/6[a]	Aoki et al, 2003[141]
	Terc	$Terc^{tm1Rdp}$	Knockout	Min	Decrease (at G4)	10	Rudolph et al, 2001[153]
DNA mutation rate	Pms2	$Pms2^{tm1Lisk}$	Knockout	Min	Increase	3	Baker et al, 1998[154]
	Mlh1	$Mlh1^{tm1Lisk}$	Knockout	Min	Increase	3	Shoemaker et al, 2000[155]
	Msh2	$Msh2^{tm1Mak}$	Knockout	Min	Increase	7	Reitmair et al, 1996[156]
	Msh3/Msh6	$Msh3^{tm1Rak}$ $Msh6^{tm1Rak}$	Knockout	1638N	Increase	12	Kuraguchi et al, 2001[157]
	Fen1	$Fen1^{tm1Rak}$	Knockout (het)	1638N	Increase	1.5	Kucherlapati et al, 2002[158]
	Myh	$Mutyh^{tm1Jhmi}$	Knockout	Min	Increase	1.5	Sieber et al, 2004[159]
Recombination rates	Rb9	Rb(7.18)9Lub	Translocation	Min	Decrease	19	Haigis and Dove, 2003[100]
	Recql4	$Recql4^{tm1Glu}$	Knockout	Min	Increase	2	Mann et al, 2005[144]
	Blm	Blm^{tm3Brd}	Hypomorph	Min	Increase	3	Luo et al, 2000[143]
Differentiation	EphB2	Δ^{cy}EphB2	Dom neg Tg	Min	Decrease	3	Batlle et al, 2005[147]
	EphB3	$EphB3^{tm1Kln}$	Knockout	Min	Increase	2	Batlle et al, 2005[147]
DNA methylation	Mbd2	$Mbd2^{tm1Bh}$	Knockout	Min	Decrease	10	Sansom et al, 2003[145]
	Mbd4	$Mbd4^{tm1Bird}$	Knockout	Min	Increase	2	Millar et al, 2002[146]
	Dnmt1	$Dnmt1^{tm1Jae}$	Knockout (het)	Min	Decrease	2	Cormier and Dove, 2000[85]

continued on next page

Table 3. Continued

Modifier Affects	Modifier Gene(s)	Modifier Allele(s)	Allele Property	Apc Model	Effect of Mutant Allele on Intestinal Tumor Multiplicity	Factor of Effect	Reference
Stromal regulation	Foxl1	$Foxl1^{tm1Khk}$	Knockout	Min	Increase	8	Perrault et al, 2005[160]
	TSP1	$Thbs1^{tm1Hyn}$	Knockout	Min	Increase	2	Gutierrez et al, 2003[161]
Cell growth and proliferation	c-Jun	$Jun^{tm2.1Wag}$	Hypomorph	Min	Decrease	2	Nateri et al, 2005[162]
	Cyclin D1	$Ccnd1^{tm1Wbg}$	Knockout	Min	Decrease	6	Hulit et al, 2004[163]
	Egfr	$Egfr^{wa2}$	Hypomorph	Min	Decrease	10	Roberts et al, 2002[164]
	p21	$Cdkn1a^{tm1Led}$	Knockout	1638N	Increase	2	Yang et al, 2001[165]
	p27	$Cdkn1b^{tm1Mlf}$	Knockout	Min	Increase	5	Philippp-Staheli et al, 2002[166]
	p53	$Trp53^{tm1Ldo}$	Knockout	Min	Increase	2	Halberg et al, 2000[167]
	Igf2	$H19^{tm1Tilg}$	Activates Ifg2	Min	Increase	2	Sakatani et al, 2005[139]
Pleiotropic	Matrilysin	$Mmp7^{tm1Lmm}$	Knockout	Min	Decrease	2	Wilson et al, 1997[168]
	Pla2g2a	$Pla2g2a^{AKR}$	Tg	Min	Decrease	2	Cormier et al, 1997[125]
	BAH	$Asph^{tm1Jed}$	Knockout	Min	Increase	2	Dinchuk et al, 2002[169]
	E-cadherin	$Cdh1^{tm1Cbm}$	Knockout (het)	1638N	Increase	9	Smits et al, 2000[170]
	PPAR-δ	$Ppard^{tm1Jps}$	Knockout	Min	Increase	1.5	Harman et al, 2004[171]
	Netrin-1	Tg-netrin-1	Tg	1638N	Enhances progression	N/A	Mazelin et al, 2004[172]
	Smad4	$Smad4^{tm1Mmt}$	Knockout	Δ716	Enhances progression	N/A	Takaku et al, 1999[173]

[a]Effects on the small intestine and colon, respectively; [b]The Robertsonian translocation is centromeric fusion of chromosomes 7 and 18. Note: The *Mom* (Modifier of Min) and *Scc* (Susceptibility to colon cancer)[174] loci are in general not yet fully defined in molecular detail (see text) and are therefore not included in Table 3.

References

1. Jemal A, Siegel R, Ward E et al. Cancer statistics. CA Cancer J Clin 2006; 56:106-130.
2. Grosch S, Maier TJ, Schiffmann S et al. Cyclooxygenase-2 (COX-2)-independent anticarcinogenic effects of selective COX-2 inhibitors. J Natl Cancer Inst 2006; 98:736-747.
3. O'Connell JB, Maggard MA, Ko CY. Colon cancer survival rates with the new american joint committee on cancer sixth edition staging. J Natl Cancer Inst 2004; 96:1420-1425.
4. Radtke F, Clevers H. Self-renewal and cancer of the gut: two sides of a coin. Science 2005; 307:1904-1909.
5. Wasan HS, Park HS, Liu KC et al. APC in the regulation of intestinal crypt fission. J Pathol 1998; 185:246-255.
6. Greaves LC, Preston SL, Tadrous PJ et al. Mitochondrial DNA mutations are established in human colonic stem cells and mutated clones expand by crypt fission. Proc Natl Acad Sci USA 2006; 103:714-719.
7. Ponder BAJ, Schmidt GH, Wilkinson MM et al. Derivation of mouse intestinal crypts from single progenitor cells. Nature 1985; 313:689-691.
8. Kim KM, Shibata D. Methylation reveals a niche: stem cell succession in human colon crypts. Oncogene 2002; 21:5441-5449.
9. Fujita Y, Cheung AT, Kieffer TJ. Harnessing the gut to treat diabetes. Pediatr Diabetes 2004; 5(Suppl 2):57-69.
10. Potten CS, Morris RJ. Epithelial stem cells in vivo. J Cell Sci Suppl 1988; 10:45-62.
11. Cheng H, Leblond CP. Origin, differentiation and renewal of the four main epithelial cell types in the mouse small intestine. V. Unitarian theory of the origin of the four epithelial cell types. Am J Anat 1974; 141:537-562.
12. Potten CS. Epithelial cell growth and differentiation. II. Intestinal apoptosis. Am J Physiol Gastrointest Liver Physiol 1997; 273:G253-G257.
13. Clarke MF, Fuller M. Stem sells and cancer: Two faces of Eve. Cell 2006; 124:1111-1115.
14. Till JE, Siminovitch L, McCulloch EA. The effect of plethora on growth and differentiation of normal hemopoietic colony-forming cells transplanted in mice of genotype W/Wv. Blood 1967; 29:102-113.
15. Reya T, Morrison SJ, Clarke MF et al. Stem cells, cancer and cancer stem cells. Nature 2001; 414:105-111.
16. Bonnet D, Dick JE. Human acute myeloid leukemia is organized as a hierarchy that originates from a primitive hematopoietic cell. Nat Med 1997; 3:730-737.
17. Al Hajj M, Wicha MS, Benito-Hernandez A et al. Prospective identification of tumorigenic breast cancer cells. Proc Natl Acad Sci USA 2003; 100:3983-3988.
18. Matsui W, Huff CA, Wang Q et al. Characterization of clonogenic multiple myeloma cells. Blood 2004; 103:2332-2336.
19. Singh SK, Clarke ID, Terasaki M et al. Identification of a cancer stem cell in human brain tumors. Cancer Res 2003; 63:5821-5828.
20. Collins AT, Berry PA, Hyde C et al. Prospective identification of tumorigenic prostate cancer stem cells. Cancer Res 2005; 65:10946-10951.
21. O'Brien CA, Pollett A, Gallinger S et al. A human colon cancer cell capable of initiating tumour growth in immunodeficient mice. Nature 2007; 445:106-110.
22. Ricci-Vitiani L, Lombardi DG, Pilozzi E et al. Identification and expansion of human colon-cancer-initiating cells. Nature 2006.
23. Polyak K, Hahn WC. Roots and stems: stem cells in cancer. Nat Med 2006; 12:296-300.
24. Shih IM, Wang TL, Traverso G et al. Top-down morphogenesis of colorectal tumors. Proc Natl Acad Sci USA 2001; 98:2640-2645.
25. Kim KM, Calabrese P, Tavare S et al. Enhanced stem cell survival in familial adenomatous polyposis. Am J Pathol 2004; 164:1369-1377.
26. Wood DA, Robbins GF, Zippin C et al. Staging of cancer of the colon and cancer of the rectum. Cancer 1979; 43:961-968.
27. Gospodarowicz MK, Miller D, Groome PA et al. The process for continuous improvement of the TNM classification. Cancer 2004; 100:1-5.
28. Bond JH. Colon Polyps and Cancer. Endoscopy 2003; 27-35.
29. Chan TL, Zhao W, Cancer GP et al. BRAF and KRAS mutations in colorectal hyperplastic polyps and serrated adenomas. Cancer Res 2003; 63:4878-4881.
30. Gardner EJ. A genetic and clinical study of intestinal polyposis, a predisposing factor for carcinoma of the colon and rectum. Am J Hum Genet 1951; 3:167-176.
31. Shoemaker AR, Gould KA, Luongo C et al. Studies of neoplasia in the Min mouse. Biochim Biophys Acta 1997; 1332:F25-F48.

32. Herrera L, Kakati S, Gibas L et al. Brief clinical report: gardner syndrome in a man with an interstitial deletion of 5q. Am J Hum Genet 1986; 25:473-476.
33. Bodmer WF, Bailey CJ, Bodmer J et al. Localization of the gene for familial adenomatous polyposis on chromosome 5. Nature 1987; 328:614-616.
34. Kinzler KW, Nilbert MC, Su LK et al. Identification of FAP locus genes from chromosome 5q21. Science 1991; 253:661-665.
35. Nishisho I, Nakamura Y, Miyoshi Y et al. Mutation of chromosome 5q21 genes in FAP and colorectal cancer patients. Science 1991; 253:665-668.
36. Joslyn G, Carlson M, Thliveris A et al. Identification of deletion mutations and three new genes at the familial polyposis locus. Cell 1991; 66:601-613.
37. Groden J, Thliveris A, Samowitz W et al. Identification and characterization of the familial adenomatous polyposis coli gene. Cell 1991; 66:589-600.
38. Fearnhead NS, Britton MP, Bodmer WF. The ABC of APC. Hum Mol Genet 2001; 10:721-733.
39. Horii A, Nakatsuru S, Miyoshi Y et al. Frequent somatic mutations of the APC gene in human pancreatic cancer. Cancer Res 1992; 52:6696-6698.
40. Clement G, Bosman FT, Fontolliet C et al. Monoallelic methylation of the APC promoter is altered in normal gastric mucosa associated with neoplastic lesions. Cancer Res 2004; 64:6867-6873.
41. Su LK, Vogelstein B, Kinzler KW. Association of the APC tumor suppressor protein with catenins. Science 1993; 262:1734-1737.
42. Rubinfeld B, Souza B, Albert I et al. Association of the APC gene product with β-catenin. Science 1993; 262:1731-1734.
43. Fearnhead NS, Wilding JL, Bodmer WF. Genetics of colorectal cancer: hereditary aspects and overview of colorectal tumorigenesis. Br Med Bull 2002; 64:27-43.
44. Noubissi FK, Elcheva I, Bhatia N et al. CRD-BP mediates stabilization of [beta]TrCP1 and c-myc mRNA in response to [beta]-catenin signalling. Nature 2006; 441:898-901.
45. Harada N, Tamai Y, Ishikawa T et al. Intestinal polyposis in mice with a dominant stable mutation of the beta-catenin gene. EMBO J 1999; 18:5931-5942.
46. Romagnolo B, Berrebi D, Saadi-Keddoucci S et al. Intestinal dysplasia and adenoma in transgenic mice after overexpression of an activated β-catenin. Cancer Res 1999; 59:3875-3879.
47. Thliveris A, Samowitz W, Matsunami N et al. Demonstration of promoter activity and alternative splicing in the region 5′ to exon 1 of the APC gene. Cancer Res 1994; 54:2991-2995.
48. Bardos J, Sulekova Z, Ballhausen WG. Novel exon connections of the brain-specific (BS) exon of the adenomatous polyposis coli gene. Int J Cancer 1997; 73:137-142.
49. Pyles RB, Santoro IM, Groden J et al. Novel protein isoforms of the APC tumor suppressor in neural tissue. Oncogene 1998; 16:77-82.
50. Rubinfeld B, Albert I, Porfiri E et al. Binding of GSK3β to the APC-β-catenin complex and regulation of complex assembly. Science 1996; 272:1023-1026.
51. Behrens J, Jerchow BA, Wurtele M et al. Functional interaction of an axin homolog, conductin, with β-catenin, APC and GSK3β. Science 1998; 280:596-599.
52. Zumbrunn J, Kinoshita K, Hyman AA et al. Binding of the adenomatous polyposis coli protein to microtubules increases microtubule stability and is regulated by GSK3[beta] phosphorylation. Curr Biol 2001; 11:44-49.
53. Polakis P. The adenomatous polyposis coli (APC) tumor suppressor. Biochim Biophys Acta 1997; 1332:F127-F148.
54. Heppner Goss K, Trzepacz C, Tuohy TM et al. Attenuated APC alleles produce functional protein from internal translation initiation. Proc Natl Acad Sci USA 2002; 99:8161-8166.
55. Jimbo T, Kawasaki Y, Koyama R et al. Identification of a link between the tumour suppressor APC and the kinesin superfamily. Nat Cell Biol 2002; 4:323-327.
56. Aoki K, Taketo MM. Adenomatous polyposis coli (APC): a multi-functional tumor suppressor gene. J Cell Sci 2007; 120:3327-3335.
57. Rowan AJ, Lamlum H, Ilyas M et al. APC mutations in sporadic colorectal tumors: A mutational "hotspot" and interdependence of the "two hits". Proc Natl Acad Sci USA 2000; 97:3352-3357.
58. Albuquerque C, Breukel C, van der LR et al. The 'just-right' signaling model: APC somatic mutations are selected based on a specific level of activation of the beta-catenin signaling cascade. Hum Mol Genet 2002; 11:1549-1560.
59. Crabtree M, Sieber OM, Lipton L et al. Refining the relation between 'first hits' and 'second hits' at the APC locus: the 'loose fit' model and evidence for differences in somatic mutation spectra among patients. Oncogene 2003; 22:4257-4265.
60. Crabtree MD, Tomlinson IPM, Hodgson SV et al. Explaining variation in familial adenomatous polyposis: relationship between genotype and phenotype and evidence for modifier genes. Gut 2002; 51:420-423.

61. Soliman AS, Bondy ML, El Badawy SA et al. Contrasting molecular pathology of colorectal carcinoma in Egyptian and Western patients. Br J Cancer 2001; 85:1037-1046.
62. Moser AR, Pitot HC, Dove WF. A dominant mutation that predisposes to multiple intestinal neoplasia in the mouse. Science 1990; 247:322-324.
63. Su LK, Kinzler KW, Vogelstein B et al. Multiple intestinal neoplasia caused by a mutation in the murine homolog of the APC gene. Science 1992; 256:668-670.
64. Sasai H, Masaki M, Wakitani K. Suppression of polypogenesis in a new mouse strain with a truncated Apc{Delta}474 by a novel COX-2 inhibitor, JTE-522. Carcinogenesis 2000; 21:953-958.
65. Colnot S, Niwa-Kawakita M, Hamard G et al. Colorectal cancers in a new mouse model of familial adenomatous polyposis: influence of genetic and environmental modifiers. Lab Invest 2004; 84:1619-1630.
66. Oshima M, Oshima H, Kitagawa K et al. Loss of Apc heterozygosity and abnormal tissue building in nascent intestinal polyps in mice carrying a truncated Apc gene. Proc Natl Acad Sci USA 1995; 92:4482-4486.
67. Ishikawa T, Tamai Y, Li Q et al. Requirement for tumor suppressor Apc in the morphogenesis of anterior and ventral mouse embryo. Dev Biol 2003; 253:230-246.
68. Niho N, Takahashi M, Kitamura T et al. Concomitant suppression of hyperlipidemia and intestinal polyp formation in Apc-deficient mice by peroxisome proliferator-activated receptor ligands. Cancer Res 2003; 63:6090-6095.
69. Fodde R, Edelmann W, Yang K et al. A targeted chain-termination mutation in the mouse Apc gene results in multiple intestinal tumors. Proc Natl Acad Sci USA 1994; 91:8969-8973.
70. Smits R, Kielman MF, Breukel C et al. Apc1638T: a mouse model delineating critical domains of the adenomatous polyposis coli protein involved in tumorigenesis and development. Genes Dev 1999; 13:1309-1321.
71. Li Q, Ishikawa TO, Oshima M et al. The threshold level of adenomatous polyposis coli protein for mouse intestinal tumorigenesis. Cancer Res 2005; 65:8622-8627.
72. Haigis KM, Hoff PD, White A et al. Tumor regionality in the mouse intestine reflects the mechanism of loss of Apc function. Proc Natl Acad Sci USA 2004; 101:9769-9773.
73. Baran AA, Silverman KA, Zeskand J et al. The modifier of Min 2 (Mom2) locus: embryonic lethality of a mutation in the Atp5a1 gene suggests a novel mechanism of polyp suppression. Genome Res 2007; 17:566-576.
74. Gounari F, Chang R, Cowan J et al. Loss of adenomatous polyposis coli gene function disrupts thymic development. Nat Immunol 2005; 6:800-809.
75. Shibata H, Toyama K, Shioya H et al. Rapid colorectal adenoma formation initiated by conditional targeting of the Apc gene. Science 1997; 278:120-123.
76. Colnot S, Decaens T, Niwa-Kawakita M et al. Liver-targeted disruption of Apc in mice activates beta-catenin signaling and leads to hepatocellular carcinomas. Proc Natl Acad Sci USA 2004; 101:17216-17221.
77. Sansom OJ, Griffiths DF, Reed KR et al. Apc deficiency predisposes to renal carcinoma in the mouse. Oncogene 2005; 24:8205-8210.
78. Sansom OJ, Reed KR, Hayes AJ et al. Loss of Apc in vivo immediately perturbs Wnt signaling, differentiation and migration. Genes Dev 2004; 18:1385-1390.
79. Andreu P, Colnot S, Godard C et al. Crypt-restricted proliferation and commitment to the Paneth cell lineage following Apc loss in the mouse intestine. Development 2005; 132:1443-1451.
80. Sansom OJ, Meniel VS, Muncan V et al. Myc deletion rescues Apc deficiency in the small intestine. Nature 2007; 446:676-679.
81. Bissahoyo A, Pearsall RS, Hanlon K et al. Azoxymethane is a genetic background-dependent colorectal tumor initiator and promoter in mice: effects of dose, route and diet. Toxicol Sci 2005; Dec;88(2):340-5. Epub 2005 Sep 8.
82. Shoemaker AR, Moser AR, Dove WF. N-ethyl-N-nitrosourea treatment of multiple intestinal neoplasia (Min) mice: age-related effects on the formation of intestinal adenomas, cystic crypts and epidermoid cysts. Cancer Res 1995; 55:4479-4485.
83. Biswas S, Chytil A, Washington K et al. Transforming growth factor beta receptor type II inactivation promotes the establishment and progression of colon cancer. Cancer Res 2004; 64:4687-4692.
84. Oshima H, Oshima M, Kobayashi M et al. Morphological and molecular processes of polyp formation in $Apc^{\Delta 716}$ knockout mice. Cancer Res 1997; 57:1644-1649.
85. Cormier RT, Dove WF. $Dnmt1^{N/+}$ reduces the net growth rate and multiplicity of intestinal adenomas in C57BL/6-Multiple intestinal neoplasia (Min)/+ mice independently of p53 but demonstrates strong synergy with the Modifier of Min 1^{AKR} resistance allele. Cancer Res 2000; 60:3965-3970.
86. Preston SL, Wong WM, Chan AO et al. Bottom-up histogenesis of colorectal adenomas: origin in the monocryptal adenoma and initial expansion by crypt fission. Cancer Res 2003; 63:3819-3825.

87. Leedham S, Wright N. Expansion of a mutated clone—from stem cell to tumour. J Clin Pathol 2008; Feb;61(2):164-71. Epub 2007 Apr 27.
88. Goodman DG, Ward JM, Squire RA et al. Neoplastic and Nonneoplastic Lesions in Aging Osborne—Mendel Rats. Toxicology and Applied Pharmacology 1980; 55:433-447.
89. Miyamoto M, Takizawa S. Colon carcinoma of highly inbred rats. J Natl Cancer Inst 1975; 55:1471-1472.
90. Corpet DE, Pierre F. How good are rodent models of carcinogenesis in predicting efficacy in humans? A systematic review and meta-analysis of colon chemoprevention in rats, mice and men. Eur J Cancer 2005; 41:1911-1922.
91. Zan Y, Haag JD, Chen KS et al. Production of knockout rats using ENU mutagenesis and a yeast-based screening assay. Nat Biotechnol 2003; 21:645-651.
92. Smits BM, Mudde JB, van de BJ et al. Generation of gene knockouts and mutant models in the laboratory rat by ENU-driven target-selected mutagenesis. Pharmacogenet Genomics 2006; 16:159-169.
93. Amos-Landgraf JM, Kwong LN, Kendziorski CM et al. A target-selected Apc-mutant rat kindred enhances the modeling of familial human colon cancer. Proc Natl Acad Sci USA 2007; 104:4036-4041.
94. Haramis AP, Hurlstone A, Van D et al. Adenomatous polyposis coli-deficient zebrafish are susceptible to digestive tract neoplasia. EMBO Rep 2006; 7:444-449.
95. McCartney BM, Price MH, Webb RL et al. Testing hypotheses for the functions of APC family proteins using null and truncation alleles in Drosophila. Development 2006; 133:2407-2418.
96. Hamada F, Murata Y, Nishida A et al. Identification and characterization of E-APC, a novel Drosophila homologue of the tumour suppressor APC. Genes to Cells 1999; 4:465-474.
97. Rocheleau CE, Downs WD, Lin R et al. Wnt Signaling and an APC-Related Gene Specify Endoderm in Early C. elegans Embryos. Cell 1997; 90:707-716.
98. Mizumoto K, Sawa H. Cortical beta-catenin and APC regulate asymmetric nuclear beta-catenin localization during asymmetric cell division in C. elegans. Dev Cell 2007; 12:287-299.
99. Luongo C, Moser AR, Gledhill S et al. Loss of Apc^{+} in intestinal adenomas from Min mice. Cancer Res 1994; 54:5947-5952.
100. Haigis KM, Dove WF. A Robertsonian translocation suppresses a somatic recombination pathway to loss of heterozygosity. Nat Genet 2003; 33:33-39.
101. Sieber OM, Heinimann K, Gorman P et al. Analysis of chromosomal instability in human colorectal adenomas with two mutational hits at APC. Proc Natl Acad Sci USA 2002; 99:16910-16915.
102. Haigis KM, Caya JG, Reichelderfer M et al. Intestinal adenomas can develop with a stable karyotype and stable microsatellites. Proc Natl Acad Sci USA 2002; 99:8927-8931.
103. Cavenee WK, Dryja TP, Phillips RA et al. Expression of recessive alleles by chromosomal mechanisms in retinoblastoma. Nature 1983; 305:779-784.
104. Hagstrom SA, Dryja TP. Mitotic recombination map of 13cen-13q14 derived from an investigation of loss of heterozygosity in retinoblastomas. Proc Natl Acad Sci USA 1999; 96:2952-2957.
105. Thiagalingam S, Laken S, Willson JK et al. Mechanisms underlying losses of heterozygosity in human colorectal cancers. Proc Natl Acad Sci USA 2001; 98:2698-2702.
106. Shih IM, Zhou W, Goodman SN et al. Evidence that genetic instability occurs at an early stage of colorectal tumorigenesis. Cancer Res 2001; 61:818-822.
107. Xinarianos G, McRonald FE, Risk JM et al. Frequent genetic and epigenetic abnormalities contribute to the deregulation of cytoglobin in nonsmall cell lung cancer. Hum Mol Genet 2006; 15:2038-2044.
108. Cardoso J, Molenaar L, de Menezes RX et al. Chromosomal instability in MYH- and APC-mutant adenomatous polyps. Cancer Res 2006; 66:2514-2519.
109. Nowak MA, Komarova NL, Sengupta A et al. The role of chromosomal instability in tumor initiation. Proc Natl Acad Sci USA 2002; 99:16226-16231.
110. Komarova NL, Wodarz D. The optimal rate of chromosome loss for the inactivation of tumor suppressor genes in cancer. Proc Natl Acad Sci USA 2004; 101:7017-7021.
111. Tomlinson IPM, Novelli MR, Bodmer WF. The mutation rate and cancer. Proc Natl Acad Sci USA 1996; 93:14800-14803.
112. Green RA, Kaplan KB. Chromosome instability in colorectal tumor cells is associated with defects in microtubule plus-end attachments caused by a dominant mutation in APC. The Journal of Cell Biology 2003; 163:949-961.
113. Tighe A, Johnson VL, Taylor SS. Truncating APC mutations have dominant effects on proliferation, spindle checkpoint control, survival and chromosome stability. J Cell Sci 2004; 117:6339-6353.
114. Li Z, Nathke IS. Tumor-associated NH2-terminal fragments are the most stable part of the adenomatous polyposis coli protein and can be regulated by interactions with COOH-terminal domains. Cancer Res 2005; 65:5195-5204.
115. Mahmoud NN, Boolbol SK, Bilinski RT et al. Apc Gene Mutation Is Associated with a Dominant-Negative Effect upon Intestinal Cell Migration. Cancer Res 1997; 57:5045-5050.

116. Carothers AM, Melstrom KA Jr, Mueller JD et al. Progressive changes in adherens junction structure during intestinal adenoma formation in Apc mutant mice. J Biol Chem 2001; 276:39094-39102.
117. Moran AE, Hunt DH, Javid SH et al. Apc deficiency is associated with increased Egfr activity in the intestinal enterocytes and adenomas of C57BL/6J-Min/+ mice. J Biol Chem 2004; 279:43261-43272.
118. Oshima M, Oshima H, Kobayashi M et al. Evidence against dominant negative mechanisms of intestinal polyp formation by Apc gene mutations. Cancer Res 1995; 55:2719-2722.
119. Sieber OM, Lamlum H, Crabtree MD et al. Whole-gene APC deletions cause classical familial adenomatous polyposis, but not attenuated polyposis or "multiple" colorectal adenomas. Proc Natl Acad Sci USA 2002; 99:2954-2958.
120. Michils G, Tejpar S, Thoelen R et al. Large deletions of the APC gene in 15% of mutation-negative patients with classical polyposis (FAP): a Belgian study. Hum Mutat 2005; 25:125-134.
121. Kwong LN, Shedlovsky A, Biehl BS et al. Identification of Mom7, a novel modifier of $Apc^{Min/+}$ on mouse Chromosome 18. Genetics 2007; 176:1237-1244.
122. Shoemaker AR, Moser AR, Midgley CA et al. A resistant genetic background leading to incomplete penetrance of intestinal neoplasia and reduced loss of heterozygosity in $Apc^{Min/+}$ mice. Proc Natl Acad Sci USA 1998; 95:10826-10831.
123. Dietrich WF, Lander ES, Smith JS et al. Genetic identification of Mom-1, a major modifier locus affecting Min-induced intestinal neoplasia in the mouse. Cell 1993; 75:631-639.
124. MacPhee M, Chepenik KP, Liddell RA et al. The secretory phospholipase A2 gene is a candidate for the Mom1 locus, a major modifier of Apc^{Min}-induced intestinal neoplasia. Cell 1995; 81:957-966.
125. Cormier RT, Hong KH, Halberg RB et al. Secretory phospholipase Pla2g2a confers resistance to intestinal tumorigenesis. Nat Genet 1997; 17:88-91.
126. Cormier RT, Bilger A, Lillich AJ et al. The $Mom1^{AKR}$ intestinal tumor resistance region consists of Pla2g2a and a locus distal to D4Mit64. Oncogene 2000; 19:3182-3192.
127. Belinsky GS, Rajan TV, Saria EA et al. Expression of secretory phospholipase A2 in colon tumor cells potentiates tumor growth. Mol Carcinog 2007; 46:106-116.
128. Dove W. Aurora and the hunt for cancer-modifying genes. Nat Genet 2003; 34:353-354.
129. Chan TA, Morin PJ, Vogelstein B et al. Mechanisms underlying nonsteroidal antiinflammatory drug-mediated apoptosis. Proc Natl Acad Sci USA 1998; 95:681-686.
130. Tomlinson IP, Beck NE, Neale K et al. Variants at the secretory phospholipase A2 (PLA2G2A) locus: analysis of associations with familial adenomatous polyposis and sporadic colorectal tumours. Ann Hum Genet 1996; 60(Pt 5):369-376.
131. Spirio LN, Kutchera W, Winstead MV et al. Three secretory phospholipase A_2 genes that map to human chromosome 1P35-36 are not mutated in individuals with attenuated adenomatous polyposis coli. Cancer Res 1996; 56:955-958.
132. Riggins GJ, Markowitz S, Wilson JK et al. Absence of secretory phospholipase A2 gene alterations in human colorectal cancer. Cancer Res 1995; 55:5184-5186.
133. Nimmrich I, Friedl W, Kruse R et al. Loss of the PLA2G2A gene in a sporadic colorectal tumor of a patient with a PLA2G2A germline mutation and absence of PLA2G2A germline alterations in patients with FAP. Hum Genet 1997; 100:345-349.
134. Leung SY, Chen X, Chu KM et al. Phospholipase A2 group IIA expression in gastric adenocarcinoma is associated with prolonged survival and less frequent metastasis. Proc Natl Acad Sci USA 2002; 99:16203-16208.
135. Silverman KA, Koratkar R, Siracusa LD et al. Identification of the modifier of Min 2 (Mom2) locus, a new mutation that influences Apc-induced intestinal neoplasia. Genome Res 2002; 12:88-97.
136. Haines J, Johnson V, Pack K et al. Genetic basis of variation in adenoma multiplicity in ApcMin/+ Mom1S mice. Proc Natl Acad Sci USA 2005; 102:2868-2873.
137. Shao C, Stambrook PJ, Tischfield JA. Mitotic recombination is suppressed by chromosomal divergence in hybrids of distantly related mouse strains. Nat Genet 2001; 28:169-172.
138. Rao CV, Yang YM, Swamy MV et al. Colonic tumorigenesis in BubR1+/-ApcMin/+ compound mutant mice is linked to premature separation of sister chromatids and enhanced genomic instability. PNAS 2005; 102:4365-4370.
139. Sakatani T, Kaneda A, Iacobuzio-Donahue CA et al. Loss of imprinting of Igf2 alters intestinal maturation and tumorigenesis in mice. Science 2005; 307:1976-1978.
140. Cui H, Cruz-Correa M, Giardiello FM et al. Loss of IGF2 imprinting: A potential marker of colorectal cancer risk. Science 2003; 299:1753-1755.
141. Aoki K, Tamai Y, Horiike S et al. Colonic polyposis caused by mTOR-mediated chromosomal instability in Apc+/Delta716 Cdx2+/− compound mutant mice. Nat Genet 2003; 35:323-330.
142. Goss KH, Risinger MA, Kordich JJ et al. Enhanced tumor formation in mice heterozygous for Blm mutation. Science 2002; 297:2051-2053.

143. Luo G, Santoro IM, McDaniel LD et al. Cancer predisposition caused by elevated mitotic recombination in Bloom mice. Nat Genet 2000; 26:424-429.
144. Mann MB, Hodges CA, Barnes E et al. Defective sister-chromatid cohesion, aneuploidy and cancer predisposition in a mouse model of type II Rothmund-Thomson syndrome. Hum Mol Genet 2005; 14:813-825.
145. Sansom OJ, Berger J, Bishop SM et al. Deficiency of Mbd2 suppresses intestinal tumorigenesis. Nat Genet 2003; 34:145-147.
146. Millar CB, Guy J, Sansom OJ et al. Enhanced CpG mutability and tumorigenesis in MBD4-deficient mice. Science 2002; 297:403-405.
147. Batlle E, Bacani J, Begthel H et al. EphB receptor activity suppresses colorectal cancer progression. Nature 2005; 435:1126-1130.
148. Leppert M, Dobbs M, Scambler P et al. The gene for familial polyposis coli maps to the long arm of chromosome 5. Science 1987; 238:1411-1413.
149. Oshima M, Oshima H, Kitagawa K et al. Loss of Apc heterozygosity and abnormal tissue building in nascent intestinal polyps in mice carrying a truncated Apc gene. Proc Natl Acad Sci USA 1995; 92:4482-4486.
150. Ishikawa TO, Jain NK, Taketo MM et al. Imaging cyclooxygenase-2 (Cox-2) gene expression in living animals with a luciferase knock-in reporter gene. Mol Imaging Biol 2006; 8:171-187.
151. Koratkar R, Silverman KA, Pequignot E et al. Analysis of reciprocal congenic lines reveals the C3H/HeJ genome to be highly resistant to Apc^{Min} intestinal tumorigenesis. Genomics 2004; 84:844-852.
152. Koratkar R, Pequignot E, Hauck WW et al. The CAST/Ei strain confers significant protection against Apc^{Min} intestinal polyps, independent of the resistant Modifier of Min 1 ($Mom1^R$) locus. Cancer Res 2002; 62:5413-5417.
153. Rudolph KL, Millard M, Bosenberg MW et al. Telomere dysfunction and evolution of intestinal carcinoma in mice and humans. Nat Genet 2001; 28:155-159.
154. Baker SM, Harris AC, Tsao JL et al. Enhanced intestinal adenomatous polyp formation in $Pms2^{-/-}$; Min mice. Cancer Res 1998; 58:1087-1089.
155. Shoemaker AR, Haigis KM, Baker SM et al. Mlh1 deficiency enhances several phenotypes of $Apc^{Min/+}$ mice. Oncogene 2000; 19:2774-2779.
156. Reitmair AH, Cai JC, Bjerknes M et al. MSH2 deficiency contributes to accelerated APC-mediated intestinal tumorigenesis. Cancer Res 1996; 56:2922-2926.
157. Kuraguchi M, Yang K, Wong E et al. The distinct spectra of tumor-associated Apc mutations in mismatch repair-deficient Apc^{1638N} mice define the roles of MSH3 and MSH6 in DNA repair and intestinal tumorigenesis. Cancer Res 2001; 61:7934-7942.
158. Kucherlapati M, Yang K, Kuraguchi M et al. Haploinsufficiency of Flap endonuclease (Fen1) leads to rapid tumor progression. PNAS 2002; 99:9924-9929.
159. Sieber OM, Howarth KM, Thirlwell C et al. Myh deficiency enhances intestinal tumorigenesis in multiple intestinal neoplasia (ApcMin/+) mice. Cancer Res 2004; 64:8876-8881.
160. Perreault N, Sackett SD, Katz JP et al. Foxl1 is a mesenchymal Modifier of Min in carcinogenesis of stomach and colon. Genes and Development 2005; 19:311-315.
161. Gutierrez LS, Suckow M, Lawler J et al. Thrombospondin 1—a regulator of adenoma growth and carcinoma progression in the APCMin/+ mouse model. Carcinogenesis 2003; 24:199-207.
162. Nateri AS, Spencer-Dene B, Behrens A. Interaction of phosphorylated c-Jun with TCF4 regulates intestinal cancer development. Nature 2005; 437:281-285.
163. Hulit J, Wang C, Li Z et al. Cyclin D1 genetic heterozygosity regulates colonic epithelial cell differentiation and tumor number in ApcMin mice. Mol Cell Biol 2004; 24:7598-7611.
164. Roberts RB, Min L, Washington MK et al. Importance of epidermal growth factor receptor signaling in establishment of adenomas and maintenance of carcinomas during intestinal tumorigenesis. Proc Natl Acad Sci USA 2002; 99:1521-1526.
165. Yang WC, Mathew J, Velcich A et al. Targeted Inactivation of the p21WAF1/cip1 gene enhances Apc-initiated tumor formation and the tumor-promoting activity of a Western-style high-risk diet by altering cell maturation in the intestinal mucosa. Cancer Res 2001; 61:565-569.
166. Philipp-Staheli J, Kim KH, Payne SR et al. Pathway-specific tumor suppression. Reduction of p27 accelerates gastrointestinal tumorigenesis in Apc mutant mice, but not in Smad3 mutant mice. Cancer Cell 2002; 1:355-368.
167. Halberg RB, Katzung DS, Hoff PD et al. Tumorigenesis in the multiple intestinal neoplasia mouse: redundancy of negative regulators and specificity of modifiers. Proc Natl Acad Sci USA 2000; 97:3461-3466.
168. Wilson CL, Heppner KJ, Labosky PA et al. Intestinal tumorigenesis is suppressed in mice lacking the metalloproteinase matrilysin. Proc Natl Acad Sci USA 1997; 94:1402-1407.

169. Dinchuk JE, Focht RJ, Kelley JA et al. Absence of posttranslational aspartyl β-hydroxylation of epidermal growth factor domains in mice leads to developmental defects and an increased incidence of intestinal neoplasia. J Biol Chem 2002; 277:12970-12977.
170. Smits R, Ruiz P, Diaz-Cano S et al. E-cadherin and adenomatous polyposis coli mutations are synergistic in intestinal tumor initiation in mice. Gastroenterology 2000; 119:1045-1053.
171. Harman FS, Nicol CJ, Marin HE et al. Peroxisome proliferator-activated receptor-[delta] attenuates colon carcinogenesis. Nat Med 2004; 10:481-483.
172. Mazelin L, Bernet A, Bonod-Bidaud C et al. Netrin-1 controls colorectal tumorigenesis by regulating apoptosis. Nature 2004; 431:80-84.
173. Takaku K, Miyoshi H, Matsunaga A et al. Gastric and duodenal polyps in Smad4 (Dpc4) knockout mice. Cancer Res 1999; 59:6113-6117.
174. Ruivenkamp CAL, Csikos T, Klous AM et al. Five new mouse susceptibility to colon cancer loci, Scc11-Scc15. Oncogene 2003; 22:7258-7260.

CHAPTER 9

Tissue-Specific Tumour Suppression by APC

Owen Sansom*

Abstract

One question that has been central to the study of the *Apc* gene is why the *Apc* gene is mutated so frequently in colorectal cancer but relatively infrequently in other tumour types. This chapter reviews recent data obtained in mice after conditional deletion of both copies of the Apc gene from adult epithelial tissues with particular focus on the intestinal epithelium. These data suggest that a major reason for the frequent mutation of *Apc* in colorectal cancer lies in the distinct character of the intestinal epithelium where Apc loss leads to a progenitor-like phenotype. Thus intestinal enterocytes lacking Apc escape the two major selective constraints that usually prevent cells from becoming cancerous in the intestine: they fail to differentiate and fail to migrate so are not sloughed off into the intestinal lumen.

Introduction

As mentioned in the chapter by Kwong and Dove, this volume, numerous insights have been gained into the role of Apc in adenoma formation from heterozygous $Apc^{Min}/+$ mice.[1] Here adenomas are formed upon loss of the remaining wild type *Apc* allele; confirming that Apc loss is the iniating event in tumour formation.[2] These mice (and other *Apc* heterozygous mice including *Apc*,[580] *Apc*,[δ716 3] *Apc*[1638N 4] mutant alleles) have been used to great effect in examining chemoprevention strategies and assessing the importance of other genes in adenoma formation (through intercrossing with other genetic mouse knockouts). However, determining the role Apc plays in the normal intestinal epithelium is difficult since the complete genetic knockout of *Apc* leads to embryonic lethality at a very early stage.[1]

To overcome the problem of embryonic lethality the bacterial CRE-LOX (Cre: Cause REcombination, LOXP1 (locus of X-over P1) system has been used in mammalian cells.[5] This system relies on a sequence specific DNA Cre recombinase which mediates intramolecular recombination and deletion of DNA between LoxP sequences. LoxP sequences are small sequences of DNA of 34 base pairs. Using gene targeting LoxP flanked alleles have been generated, allowing deletion of genes upon expression of Cre recombinase. Shibata et al[6] produced the LoxP flanked *Apc* allele 580S, which leads to the production of a truncated *Apc* allele following Cre recombinase expression. To permit gene deletion in the tissue of interest, tissue specific *Cre* transgenes have been generated so that currently genes can be deleted in nearly every tissue of the mouse.[7] This has led to new insights into the role Apc of in development, homeostasis and tumorigenesis in a number of different tissues, which has helped to explain why Apc acts a 'gatekeeper' against tumourigenesis specifically in colon.

*Owen Sansom—Beaton Institute of Cancer Research, Glasgow. G61 1BD. UK
Email: o.sansom@beatson.gla.ac.uk.

APC Proteins, edited by Inke S. Näthke and Brooke M. McCartney.

Apc *Gene Deletion in the Intestinal Epithelium*

Cell renewal within the intestinal epithelium is highly regulated and position dependent. Absorptive cells generated within intestinal crypts migrate upwards until they are shed into the gut lumen; the entire process from cells' "birth" to "death" takes 3-5 days.[8] Within the intestinal crypt there are thought to be 4-6 slowly dividing stem cells that give rise to rapidly dividing transit amplifying cells that differentiate into absorptive cells within the intestine.[9] Other cell types that are ultimately produced by the stem cells are the mucin secreting goblet cells, the hormone secreting enteroendocrine cells and the downwardly migrating, lysozyme-secreting antibacterial paneth cells (see Figure 1 for overview of intestinal epithelium.)

Given the high turnover of the intestinal epithelium, it is necessary to express Cre-recombinase within the intestinal stem cells to permanently delete genes containing Lox-P sites. Two inducible Cre transgenes with this property exist: the Aryl Hydrocarbon AhCre[10] and the Villin CreER transgene.[11] The AhCre transgenic mouse uses the inducible Cytochrome P4501A1 (CYP1A1) promoter to drive Cre expression. Cre expression is induced by injection of an inducer of cytochrome p450 such as β-napthoflavone. This small molecule binds the Ah receptor at the cell surface to cause its translocation into the nucleus to bind xenobiotic binding elements in the CYP1A1 promoter and induce cre expression.[10,12] This leads to high levels of inducible Cre expression in nearly all small intestinal enterocytes and liver hepatocytes, but also in the colon albeit at lower levels. The villin Cre ER transgene utlises the villin promoter to drive Cre expression in intestinal enterocytes. This Cre is inactive under normal circumstances; however, injection of 4-Hydroxytamoxifen (4OHT) causes the activation and expression of the villin-cre, predominantly in the small but also in the large intestine.

Both of these inducible Cre transgenic mice have been intercrossed with mice carrying loxp flanked *Apc*^{580s} alleles to have produce similar results.[11,13] Deletion of *Apc* from the intestinal crypt produced a rapid and marked phenotype (see Fig. 1): within 4 days after loss, intestinal crypts were greatly enlarged, with intestinal enterocytes proliferating independently of position. However, this was accompanied by an increase in apoptosis. Differentiation was affected as indicated by a marked reduction in goblet, enteroendocrine and absorptive cells. Importantly, Apc-deficient enterocytes also stopped migrating along the crypt-villus axis. We can thus describe cells that lack Apc in intestinal crypts as crypt progenitor cell-like.[13] Therefore Apc loss overcomes two important factors that normally constrain cancer cells in the intestine: continued proliferation and lack of elimination by migration and sloughing off.

Loss of Apc outside the intestinal crypt appears to have a distinct phenotype. Using villin Cre-ER, *Apc* was deleted from the villus compartment to produce cells that appeared relatively normal.[11] One of the most intriguing populations of cells to examine the consequence of Apc loss is where cells just exit the crypt to join the villus. Genes can be deleted in this region using both of the Cre transgenes described above. Upon depleting Apc, cells in this region appear to dedifferentiate back to crypt-like cells and lack expression of villi markers. However, they differ from crypt cells that lack Apc in that they have larger nuclei, appear polyploid and have increased levels of Cyclin D1 and p21.[11,14,15] Thus even within the intestinal epithelium the phenotype that is produced by deleting *Apc* depends on the cellular context (Fig. 1 summarises this data).

The Phenotype of *Apc* Loss and Wnt Signalling

As has been mentioned in previous chapters, one of the key functions of the Apc gene is to act as a negative regulator of Wnt signalling where it is required for the turnover of β-catenin.[16] Following Apc loss, β-catenin is no longer phosphorylated by GSK3 and so it is no longer targeted for degradation by the proteosome. Instead β-catenin now translocates to the nucleus where it interacts with LEF/TCF transcription factors to activate Wnt target genes. Consistent with these in vitro reports, the onset of the crypt progenitor cell like phenotype following *Apc* gene deletion correlates with nuclear accumulation of β-catenin.[13] A definitive experiment to show whether the accumulation of β-catenin is sufficient for all the phenotypes produced by *Apc* loss would be to delete both β-catenin and Apc simultaneously and thus ascertain the Wnt-dependent and independent functions of Apc loss. However, conditional deletion of β-catenin from the mouse intestine is highly toxic to crypt enterocytes and all β-catenin-deficient crypt enterocytes are lost 4 days after gene deletion, making

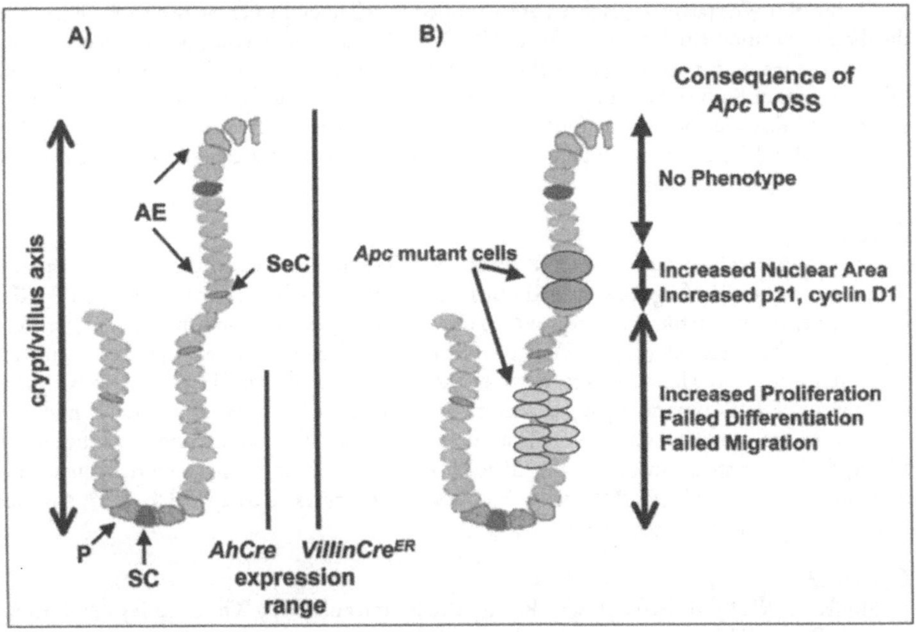

Figure 1. The position of Apc loss within the crypt is important for the precise phenotypic outcome. A) Expression of the intestinal AhCre and VillinCreER transgenes. Diagram showing simplistic version of the intestinal epithelium. The crypt (C) contains the stem cells (SC) and paneth cells (P) at its base. The rest of the crypt is composed of rapid dividing intestinal enterocytes and the secretory lineages marked in blue which are either goblet cell or enteroenodcrine cells (SeC). The Cells migrate onto the villus (V) which is composed predominanlty of absortptive enterocytes (AE) which are then sloughed off into the intestinal lumen. The red cell in the villus represents a cell which has originally produced by the red epithelial stem cell and moved onto the villus. Using the AhCre transgene, following the intraperitineal injection of a cytochrome p450 inducer, the Cre is induced within the intestinal crypt (though not the villus). Using the VillinCreEr transgene following the intraperitineal injection of tamoxifen, Cre is induced througout the entire crypt villus axis. In both cases Cre expression occurs within the intestinal stem cells so gene deletion is permenent. B) Phenotype of *Apc* loss in different populations of cells within the intestine. Using both Cre recombinases, when *Apc* is deleted within the crypt (yellow cells) this leads to a rapid induction of proliferation and the induction of a crypt progenitor cell-like phenotype. When *Apc* is deleted at the crypt-villus junction (green cells), cells upregulate p21 and Cyclin D1 and become polyploid (and are retained). When *Apc* is lost within the villus using the villin Cre Er transgene despite nuclear β-catenin, there is no observable phenotype. A color version of this image is available at www.landesbioscience.com/curie.

this experiment impossible.[10] This is probably due to the requirement for Wnt signalling within the intestinal crypt as mice ectopically expressing the Wnt repressor Dikkopf causes a large reduction in intestinal proliferation and leads to crypt death.[17]

An alternative, more feasible approach is to identify Wnt target genes that are upregulated in the intestine following Apc loss in vivo and assess the importance of these genes for the increased proliferation, reduced migration and differentiation observed. One way to identify these genes in vivo is to compare the transcriptome profile by microarray analysis before and after Apc loss.[13,18] Interestingly, despite nuclear β-catenin in every cell only a small proportion of the published Wnt target genes are upregulated in intestinal tissue following Apc loss (around 50/300). This probably reflects the differences between in vivo and in vitro experiments in cancer cell lines where most Wnt target genes were identified.[14]

Of the 300 Wnt target genes that were identified in cell lines, I will now review the impact of the three of the most cited examples, PPARβ/δ, Cyclin D1 and Myc on the phenotypes of *Apc* loss in vivo. The reason for choosing these three genes is the impact of these genes have been assessed following loss of both copies of *Apc* in intestinal cells in vivo. Thus the precise function of these genes can be dissected out in vivo. Other well known Wnt target genes such as the EphB receptors have been also shown to be important in vivo but appear to be downstream of Myc and so will be discussed in the Myc section.

PPARβ/δ

Peroxisome proliferator-activated receptors (PPARs) are members of the nuclear hormone receptor superfamily of ligand-activated transcription factors. PPARβ (also known as PPARδ) heterodimerizes with retinoic X receptor α to form a transcriptionally competent complex, that is expressed in the intestine[19] (Fig. 2). PPARβ/δ was first identified in a screen for genes upregulated in colorectal cancer cell lines and is repressed by the re-expression of *Apc*.[20] However, following *Apc* loss in vivo, *PPARβ/δ* was not upregulated nor was it upregulated in polyps from $Apc^{Min/+}$ mice; on the contrary, its expression was reduced in adenomatous tissue.[19,21] Most importantly, knockout of PPARβ/δ did not affect any of the immediate phenotypes of *Apc* loss. Furthermore, spontaneous tumourigenesis in the $PPAR\beta/\delta^{-/-}$ $Apc^{Min/+}$ mouse was not suppressed compared to $Apc^{Min/+}$, in fact a slight acceleration was detected.[19,21]

Cyclin D1

Similar to *PPARβ/δ*, *Cyclin D1* was also found to be upregulated by Wnt signalling in colorectal cancer cell lines.[22] However, although it the most cited Wnt target gene and has been postulated to be essential for Wnt mediated proliferation (Fig. 2), *Cyclin D1* is only upregulated in a small subset of cells immediately following *Apc* loss in vivo.[14] This suggests that nuclear β-catenin is not

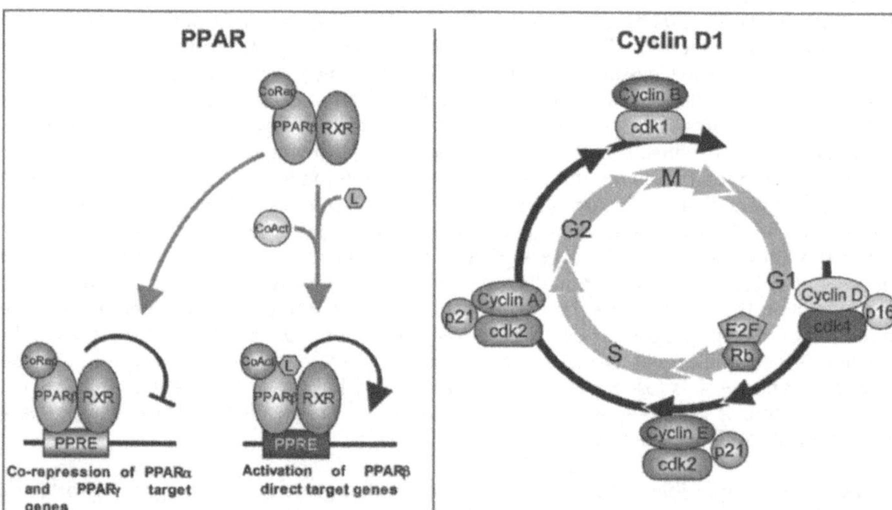

Figure 2. Overview of PPARβ/δ and Cyclin D/CDK4 pathways. A) PPARβ/δ signalling pathway. PPARβ/δ heterodimerizes with retinoic X receptor α (RXR) to form a transcriptionally competent complex. PPARβ/δ interacts with RXR and transcriptional corepressors (Co-Rep) to cause repression of PPARα/γ target genes. On ligand binding (L), PPARβ/δ interacts with RXR and transcriptional co-activators (Co-Act) to cause transcriptional activation of PPARβ/δ target genes. B) CyclinD/CDK cell cycle. D-type cyclins are important mediators of the cell cycle. The interactions of D-types cyclins with Cyclin Dependent Kinases (CDK) 2/4/6 control the G1/S-phase transition. They do this via phosphorylation and inactivation of tumor suppressor of retinoblastoma family.

sufficient to induce Cyclin D1 in intestinal epithelium. This conclusion is consistent with recent data suggesting that activation of *Cyclin D1* requires both Wnt and TGFβ signalling in the intestine.[23] Indeed, using the same colorectal cancer cell lines used by Tetsu and McCormick 1999, Sansom et al 2005 showed that the Cyclin D1 luciferase reporter was responsive to Wnt inhibition, whereas endogenous levels of Cyclin D1 were not affected. Given that Cyclin D1 was not upregulated immediately following *Apc* loss in vivo it was not surprising that knockout of *Cyclin D1* had no effect on proliferation, differentiation or migration following *Apc* loss.[14] Therefore, other target genes must be involved in the increased proliferation following *Apc* loss.

Myc

The proto-oncogene Myc has been implicated in stem cell survival, proliferation, apoptosis and tumourigenesis within a number of tissues.[24,25] Myc (c-Myc) is a transcriptional activator and repressor and is a key oncogene commonly upregulated in cancer. Myc is upregulated co-incidentally with nuclear β-catenin following *Apc* loss.[13] Complete genetic knockout of *Myc* causes embryonic lethality.[26] Thus, to examine the significance of *Myc* upregulation following *Apc* loss, the importance of Myc for intestinal homeostasis had to be examined first.[27,28] Deletion of *Myc* has varying affects on intestinal homeostasis: Bettess et al[27] found that, although Myc was important for the developing intestine, it was not required for normal intestinal homeostasis. On the other hand, Muncan et al[28] showed that the immediate consequences of *Myc* deletion included reductions in cell size, metabolism and proliferation. Over the longer term (2-4 weeks), this led to loss of Myc-deficient cells from the intestinal tissue. The reasons for the observed differences are unclear, but differences in the Cre transgenes and genetic backgrounds are likely to blame. Common to both reports was the finding that Myc-deficient cells were viable (albeit over a short term). This allowed us to determine the impact of *Myc* deletion on *Apc*-deficient cells. Mice carrying the AhCre transgene were intercrossed to mice bearing LoxP flanked *Apc* and *Myc* floxed knockout alleles.[29] Remarkably, deletion of *Myc* rescued virtually all of the phenotypes of *Apc* loss: 4 days following *Apc* loss, crypts were no longer increased in size and intestinal enterocytes migrated, proliferated and differentiated as normal wild type cells.[29] This was despite the fact that all intestinal enterocytes still had nuclear β-catenin, indicating that β-catenin was not sufficient to impose the crypt progenitor phenotype in the absence of Myc. Because Myc can modulate transcription, it was important to examine the transcriptome in intestines lacking both *Apc* and *Myc*. Almost half of the Wnt target genes that were deregulated following *Apc* loss were no longer upregulated. Furthrmore, genes that are normally repressed by Myc, like p21, were now upregulated. This suggests Myc acts a central node downstream of β-catenin upregulation to control the transcriptional activation and repression of genes that are required to maintain *Apc*-deficient cells in a crypt progenitor cell like phenotype (Fig. 3 presents a model of how Myc acts a central node for Wnt signalling).

One of the best examples of how the molecular changes following *Apc* loss (which are reliant on Myc expression) can explain phenotypic changes relates to paneth cell positioning and the expression pattern of EphB receptors. EphB2 and EphB3 receptors control the positioning of paneth cells within the normal crypt. They are expressed at the base, the site with the the highest level of Wnt signalling.[30] Following *Apc* loss, β-catenin is expressed throughout the crypt, which causes the upregulation of the EphB2 and EphB3 receptors and so removes the gradient of EphB expression that normally positions paneth cells.[13] Following the combined loss of *Apc* and *Myc*, EphB receptors are no longer upregulated (they require Myc) and paneth cells remain positioned at the base of the crypt.[29] It is possible that the failure of Apc-deficient cells to migrate along the crypt-villus axis is caused by increased levels of the pro-adhesive *TIAM1* (Tumour Invasion and Metastasis 1) gene, which is also dependent on Myc. Recent studies showed that loss of TIAM1 slows tumour formation in the $Apc^{Min/+}$ mouse but the tumours that arise are more invasive.[31] However it cannot be ruled out that the migration changes may simply represent a change in cell fate. Future studies are required to identify other Wnt targets that are upregulated in Myc-dependent manner to determine whether they are responsible for changes induced by *Apc* loss. Such an approach could help resolve whether Myc-dependent upregulation of CDK4 is essential for increased

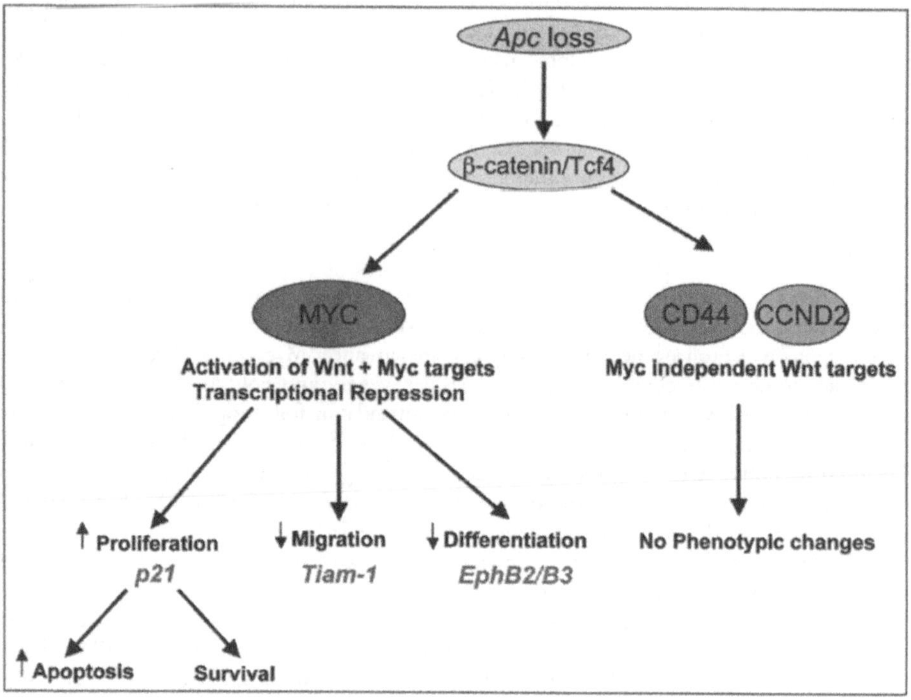

Figure 3. Myc is essential for the phenotypes of *Apc* loss. Model showing the importance of Myc for the phenotypes of *Apc* loss. Following *Apc* loss there is an induction of nuclear β-catenin which results in an upregulaltion of two sets of Wnt target genes; Myc dependent or Myc independent. Despite nuclear β-catenin the Myc independent Wnt target genes are not sufficient to induce any of the phenotypes of *Apc* loss. Further study of the Myc dependent Wnt target genes should allow us to dissect out which of these genes are important for the the phenotypes of *Apc* loss namely proliferation, differentaition and apoptosis. Abbreviations: Putative candidate genes EphB2/3: differentiation; TIAM1: migration; and p21 repression/CDK4 activation: proliferation; are highlighted on the figure.

proliferation. Targets that are required for the phenotype(s) induced by *Apc* loss may represent important therapeutic targets for colorectal cancer.

Wnt-Independent Consequences Induced by Loss of *Apc*

As has been mentioned in previous chapters, in addition to controlling Wnt signalling Apc plays Wnt-independent roles in cytoskeleton and mitosis.[15,32,33] The finding that Myc loss can rescue many of the phenotypes of *Apc* loss in vivo raises questions about the importance of these other functions of Apc. In this context, it is important to emphasise that although Myc is necessary for the phenotypes produced by *Apc* loss, it has not be shown to be sufficient for this effect. Thus other functions of Apc in addition to other Wnt targets could be crucial for colorectal carcinogenesis. For example, the large increase in proliferation that is induced by *Apc* loss could expose the importance of Wnt-independent effects of Apc in mitosis and migration. Without Myc, there will be no increase in proliferation, so that the defects in mitosis that result from *Apc* loss (due to Wnt independent functions) do not have a chance to support the genomic instability that are usually observed at later stages of colorectal cancer.[15,34]

One of the difficulties in assessing these additional functions of Apc is that, due to the difficulty in overexpressing full length Apc protein, many approaches have relied on overexpressing portions of Apc carry point mutations. The general problems associated with overexpression studies and the difficulty introduced by the fact that a number of distinct domains in the Apc molecule that are often separated by long stretches in the linear sequence are involved in specific functions make it difficult to assess the significance of these other functions independently of changes in Wnt signalling. One way to overcome these problems is to make knock-in point mutations of *Apc* using the endogenous allele (Fig. 4, *Apc* point mutations on the protein structure). Thus far only one allele has been produced that can distinguish some of the Wnt-independent functions of Apc.[35] This allele is a truncation allele position 1638, *Apc1638T*, that, unlike all the other *Apc* alleles used so far, contains the first conductin/axin binding SAMP repeat and retains the ability to regulate β-catenin. Analysis of this allele found normal levels of Wnt signalling and TCF/LEF transcription suggesting that any affects produced by this allele are Wnt-independent. Homozygous *Apc*$^{1638T/1638T}$ mice are viable, but they have a number of defects. They lack preputial glands, develop cutaneous cysts and are growth retarded. In terms of cancer predispostion *Apc*$^{1638T/1638T}$ mice are not tumour prone, arguing that removing the C-terminal domain of Apc is not sufficient to induce tumourigenesis. Attempts to generate compound heterozygotes by crossing Apc$^{1638T/1638T}$ with mice that have a loss of function *Apc* allele like *Apc*$^{1638T/Min}$ or *Apc*$^{1638T/1638N}$ led to embryonic lethality (see figure 4 for the structure of these *Apc* alleles). Future studies using the conditional LoxP alleles of *Apc* together with the *Apc*^{1638T} crossed to intestinal Cre-transgenes will allow comparisons to *Apc*$^{fl/+}$ mice and will be pivotal in examining whether the *Apc*$^{580s/1638T}$ mice show increased adenoma formation and/or whether the *Apc*^{1638T} allele is lost in adenoma.

One reason for the debate over the Wnt-independent functions of Apc for colorectal carcinogenesis, is that two loss-of-functions mutations are required in *Apc* whereas a single activating mutation in β-catenin (which prevents its phosphorylation by GSK3 and thus its targeting for degradation) is sufficient to activate Wnt target genes. Therefore given that one scenario requires 2 mutations in the same gene in the same cell (or stem cell lineage) whilst other require a single mutation, one would predict that activating *β-catenin* mutations should be much more prevalent (however they are very rare in colorectal cancer). One potential explanation is that LOF mutations occur in many

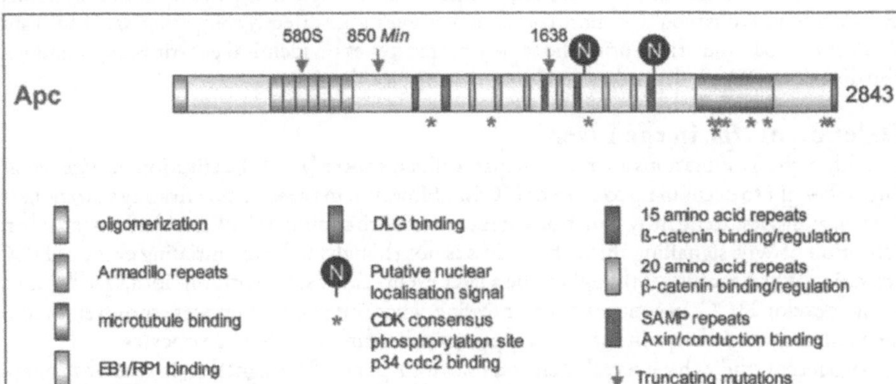

Figure 4. Apc truncation mutants. Figure showing the structure of the *Apc* gene and the *Apc1638T* allele. The arrows in red mark where the *Apc580s*, *Apc Min* and *Apc1638* mutations are in the *Apc* alleles. Briefly, the *Apc 1638T* allele still contains the first axin/conductin binding repeat which allows it to bind to the destruction complex and turnover β-catenin. Although identical the *Apc1638N* allele contains a neomycin cassette in the opposite orientation to the *Apc* gene locus, which interferes with the transcription of this allele so that protein from this the truncated allele cannot be detected. Thus this mutant cannot bind to the destruction complex and is equivalent to a knock out allele of *Apc*. Comparison of these two alleles can thus dissect out the impact of the Wnt independent functions of the Apc protein. A color version of this image is available at www.landesbioscience.com/curie.

different sites of the *Apc* gene and may represent a larger and thus more frequently hit target for mutations than a single site in *β-catenin*. However, in other cancers such as hepatocellular cancer, *Apc* mutations are rare and activating mutations of *β-catenin* are more frequent.[36] This would argue that a direct comparison between a knock-in, activated allele of *β-catenin* $Catnb^{+/lox(ex3)}$ (where the phosphorylation domain of *β-catenin* is flanked by Loxp sites so that after induction of Cre Recombinase β-catenin could no longer be targeted for degradation by the proteosome) with a knockout of both alleles of *Apc* could reveal whether loss of both copies of *Apc* provides a significant advantage for intestinal enterocytes over gaining one copy of constitutively active β-catenin. A direct comparison of these alleles using the same inducibe, intestinal cre-recombinase promoters has not yet been published. The effect of activating mutations of *β-catenin* in the intestine was examined using Keratin 5Cre to drive Cre expression. *K5Cre* $Catnb^{+/lox(ex3)}$ mice rapidly developed intestinal polyposis, with adenomas comparable to those in the $Apc^{Min/+}$ mouse. Preliminary studies in our own laboratory using the AhCre to drive *β-catenin* (*AhCre* $Catnb^{+/lox(ex3)}$) expression in the intestine suggest that the onset of the intestinal phenotype is slower compared to the situation when both copies of *Apc* were deleted. When *Apc* is deleted enlarged crypts are observed from day 4 and with increased levels of nuclear β-catenin and transcription of Wnt target genes such as Myc. However in *AhCre* $Catnb^{+/lox(ex3)}$ mice it takes longer before a phenotype is detectable (over 7 days) and also longer time before increased levels of nuclear β-catenin are observed. One possible explanation for this difference is that complete loss of *Apc* prevents all targeting of β-catenin for degradation so that it accumulates rapidly. On the other hand, following a single activating mutation in *β-catenin*, 50% of the β-catenin can be still targeted for degradation and so it takes longer to accumulate. Based on the fact that intestinal cells turn over every 3-5 days, this data argues that loss of *Apc* from cells anywhere within the crypt will allow these cells to persist and form adenomas. However, unless cells acquire an activating *β-catenin* mutation while still in the stem cell compartment, they would differentiate and be eliminated before they can alter their phenotype. This idea is speculative, but could explain the high frequency of colorectal cancers that have *Apc* mutations rather than activating *β-catenin* mutation, as the *Apc* mutations are retained within the intestinal epithelium whilst the *β-catenin* mutations are lost. This data would also predict that in other tissues with less turnover, mutations in other components of the Wnt signalling pathway would be retained in the tissue and thus would be more common. Future studies directly comparing *Apc* (LOF) and *β-catenin* (GOF) mutations using the same Cre transgenes on identical genetic background will also dissect any Wnt independent functions that Apc loss has in vivo.

Deletion of *Apc* in the Liver

Although *Apc* mutations are rare in hepatocellular cancer (HCC), activation of Wnt signalling is thought to occur in up to a 1/3 of HCC.[36] However, in these cases activating mutations in *β-catenin* and less frequently mutations in *Axin* are more common. Unlike colon cancer, where activation of Wnt signalling through *Apc* loss is not thought to be an initiating event of HCC activating Wnt signalling is thought to be a later event. Consistent with this idea, $Apc^{Min/+}$ mice do not develop HCC. In contrast to the intestine, where intestinal enterocyte turnover is rapid, the predominant cell type in the liver are postmitotic, differentiated hepatocytes.

Two recent studies have revealed critical roles for Apc and Wnt signalling in liver physiology: first, Apc was shown to have a physiological role in the liver and acts as a zonation gate keeper;[37] second, Wnt signalling was shown to be important for the massive burst of proliferation during liver regeneration after injury.[34] During regeneration, β-catenin relocalises to the nucleus and activates Wnt target gene expression. More importantly, this has been show to be functionally important as deletion of *β-catenin* (using a liver specific Cre and a Loxp flanked *β-catenin* alleles) prevents regeneration following injury.[38]

Within the liver, deletion of *Apc* or activation of β-catenin results in similar phenotypes. Both cases produce a massive burst of proliferation (akin to regeneration) and hepatomegally, i.e., the liver doubles in size.[39,40] If *Apc* is depleted or β-catenin is activated in a high percentage of hepatocytes, liver failure ensues before tumourigenesis occurs. If *Apc* is lost or *β-catenin* is activated in a smaller subset of hepatocytes (though still in around 10%), HCC is the result, although this is infrequent and requires additional genetic mutations to produce cancer such as *H-ras* activating mutations or *p53* inactivating mutations.[39-41] If we compare this to colorectal cancer, although Ras and P53 mutations accelerate intestinal tumour progression, it appears Apc loss alone is sufficient to induce tumour formation. It is important to note that Myc is required for neither the regeneration following liver injury[42] nor the proliferation that results from *Apc* loss (personal data). This highlights that the cellular context is essential for the downstream outcome of an *Apc* mutation.

Deletion of *Apc* in the Kidney

Unlike colon cancer and HCC, where a large amount of genetic evidence exists for the importance of mutations of tumour suppressor and activation of oncogenic pathways, very little is known about the molecular genetics of renal carcinoma. Loss of heterozygosity at the *Apc* locus has been reported in up to 50% of renal carcinomas. However, it is still unclear whether this due to loss of *Apc* or a nearby gene.[43] A further line of evidence suggesting that Wnt signalling may be deregulated in renal cancer is the fact that nuclear β-catenin has been reported in a significant proportion of renal cancers.[44] It is possible that in kidney cancer, mutations in other members of the Wnt pathway contribute, but these have yet to be described. Two studies investigated whether Apc acts as a tumour suppressor in the kidney using deletion of *Apc* in renal epithelium of the mouse.[45,46] One used an approach where *Apc* was deleted throughout the entire kidney using a constitutively active kidney-specific cadherin (*Ksp-Cre*) Cre recombinase. The other approach used the AhCre mouse, which has spontaneous sporadic recombination within the kidney. Kidneys from *AhCre*$^{+}$ *Apc*$^{580s/580s}$ appeared normal, although mice begun to develop renal carcinoma from 6 months of age. However, only 30% of these mice developed renal lesions by 9 months.[46] To put this into perspective, intestinal adenoma formation occurs in all mice 20 days following Apc loss following intestinal Cre induction in these *AhCre*$^{+}$ *Apc*$^{580s/580s}$ mice. When *AhCre*$^{+}$ *Apc*$^{580s/580s}$ mice were intercrossed with either *p53* knockout mice or mice that carry an oncogenic LXL (Lox Stop Lox) *KrasV12D* mutation (*AhCre*$^{+}$ *Apc*$^{580s/580}$*K-ras*$^{V12D/+}$), renal carcinoma were rapidly accelerated so that mice had kidney tumours by 2 months of ages.[47] Unlike Apc deficient mice which look histologically normal, kidneys from *AhCre*$^{+}$ *Apc*$^{580s/580s}$ *p53*$^{-/-}$ or *AhCre*$^{+}$ *Apc*$^{580s/580s}$ *K-ras*$^{V12D/+}$ also had 100's of small lesions by 2 months of age. This raised the possibility that *Apc* loss alone within the renal epithelium confers a selective disadvantage and that most mutant cells are lost and replaced wild type cells. However in the presences of oncogenic mutation (KRAS) or loss of tumour suppressor (p53), these allow the survival of these Apc deficient cells which can then go on to form tumours. To test this idea, the level of recombination in the kidney was assayed using the *LXL ROSA26R lacZ* reporter allele for recombination to assess whether cells there is a reduction in the levels of this reporter in *AhCre*$^{+}$ *Rosa26*$^{fl/+}$ Apc$^{580s/580s}$ when compared to 'wild type' *AhCre*$^{+}$ *Rosa26*$^{fl/+}$ *Apc*$^{+/+}$ kidneys. As predicted, *Apc*-deficient *AhCre*$^{+}$ *Apc*$^{580s/580s}$ kidneys showed a dramatic reduction in the number of recombined cells compared to wild type *AhCre*$^{+}$ *Apc*$^{+/+}$ kidneys.[46] This argues that within the kidney, lack of *Apc* causes cells to be deleted so that in the *AhCre*$^{+}$ *Apc*$^{580s/580s}$ kidneys only a small proportion of *Apc* depleted cells remain, which rarely progress to tumourigenesis. However, when *Apc* deficiency is combined with *p53* mutations or activating *K-ras* mutations, cells persist and tumour development is rapid. This scenario is distinctly different from the phenotype within the intestine, where loss of *Apc* is sufficient to confer a selective advantage on cells. It is also interesting to note that neither loss of p53 nor K-ras activation affect the early phenotypes produces by *Apc* loss in the intestine.[47] Therefore this data argues that whilst activation of Wnt signalling with renal epithelium is unlikely to iniate tumourigenesis, it would collaborate with other oncogenic or tumour suppressor mutations.

Deletion of *Apc* in the Mammary

There is some debate over the frequency and the significance of *Apc* loss in breast cancer. Whilst there are few reports for *Apc* mutations, hypermethylation of the *Apc* promoter has been found in 30-40% of primary breast cancers.[48,49]

To test whether *Apc* loss is sufficient to induce mammary tumourigenesis *Apc*$^{580s/580s}$ mice have been crossed to mice carrying a mammary specific β-lactoglobulin Cre transgene (*BLG Cre*).[50,51] In a situation analogous to the kidney, in the developing mammary, *Apc*-deficient cells were lost, leading to the slow growth of mammary epithelium from nonrecombined, wild type cells.[51] To induce Cre-mediated deletion of *Apc* at later stages during mammary development, female mice were mated, to produce high levels of recombination during pregnancy and lactation. Mice lacking Apc in the mammary epithelium suckled pups poorly, as indicated by slowed weight gain of pups. Histological examination of the mammary epithelium revealed transdifferentiation into squamous metaplasia.[51] As expected these cells showed high levels of β-catenin. Importantly, even when these mice were aged to 18 months, they were not prone to mammary tumours. Therefore, *Apc* loss alone was not sufficient to induce tumourigenesis in the mammary gland. The precise level of Wnt signalling does appear to be important for tumour initiation in the mammary gland. Mice highly overexpressing Wnt1 develop mammary tumours and when when *BLG-Cre Apc*$^{580s/580s}$ mice were intercrossed to mice with mice deficient for TCF1 (a transcriptional repressor of canonical TCF/LEF transcription), all mice developed multiple mammary tumours by 6 weeks of age. Thus very high levels of Wnt signalling induce tumourigenesis whilst lower levels induced by *Apc* loss induce differentiation. In a scenario analogous to the kidney and the liver, intercrossing *BLG-Cre Apc*$^{580s/580s}$ mice with *p53* knockout mice led to tumour formation.[52]

Conclusion

Comparing the phenotypes produced by Apc loss in a range of different epithelia has highlighted why Apc is such a key tumour suppressor in intestinal epithelium. Within the intestinal epithelium, Apc loss alone is sufficient to induce a crypt progenitor cell like fate and induces the onset of benign tumour formation. The escape from differentiation and migration gives Apc-deficient cells a selective advantage and allows them to persist and acquire other mutations such *K-ras*^{V12}, *p53*, which are required for later stages of colorectal carcinogenesis.[47] Within other tissues, such as kidney or mammary, loss of Apc produces a selective disadvantage so that cells are deleted or resistant to transformation following Apc loss and other mutations are required for tumourigenesis. This does not preclude the Wnt signalling pathway as an important factor for tumourigenesis in these other tissues but it does exclude *Apc* mutation as the initiating event. In these other tissues, the amplitude of wnt signalling may be important, for example in the case of the mammary gland, an extra feedback loop (TCF1) must be lost together with Apc before mammary tumours can be formed. However, within all other epithelial tissues studied so far, activation of Wnt signalling co-operates with other oncogenic mutations such as *p53* loss or *K-ras* activation. This argues that Wnt signalling may be deregulated commonly in these tumour types. Indeed, the numbers of tumours with activated Wnt signalling may be far higher than reported as mutations in other negative regulatory components or the overexpression of Wnt ligands could also contribute to dysregulated Wnt signalling. The routine investigation into these alternative modulators of Wnt signalling is just beginning.

In conclusion, the response of the intestinal epithelium to Apc loss appears to be a binary switch, so that as soon as Apc is lost cells are locked in a progenitor cell-like state. It is interesting to note that drugs like aspirin that are strong chemopreventitives not only inhibit COX2, but also can downregulate TCF/Lef signalling.[53] Studies of these drugs in the *Apc*$^{Min/+}$ mouse have highlighted that these agents stimulate migration and reduce proliferation thus overcoming the two most important selective advantage of *Apc* mutant cells.[54,55] Further work to dissect the important downstream effectors of Apc loss may allow us to increase not only our knowledge of colorectal tumourigenesis but may also aid in our ability to prevent and treat this disease.

References

1. Su LK, Kinzler KW, Vogelstein B et al. Multiple intestinal neoplasia caused by a mutation in the murine homolog of the APC gene. Science 1992; 256(5057):668-670.
2. Luongo C, Moser AR, Gledhill S et al. Loss of Apc+ in intestinal adenomas from Min mice. Cancer Res 1994; 54(22):5947-5952.
3. Oshima M, Dinchuk JE, Kargman SL et al. Suppression of intestinal polyposis in Apc delta716 knockout mice by inhibition of cyclooxygenase 2 (COX-2). Cell 1996; 87(5):803-809.
4. Yang K, Edelmann W, Fan K et al. A mouse model of human familial adenomatous polyposis. J Exp Zool 1997; 277(3):245-254.
5. Sauer B, Henderson N. Site-specific DNA recombination in mammalian cells by the Cre recombinase of bacteriophage P1. Proc Natl Acad Sci USA 1988; 85(14):5166-5170.
6. Shibata H, Toyama K, Shioya H et al. Rapid colorectal adenoma formation initiated by conditional targeting of the Apc gene. Science 1997; 278(5335):120-123.
7. Soriano P. Generalized lacZ expression with the ROSA26 Cre reporter strain. Nat Genet 1999; 21(1):70-71.
8. Marshman E, Booth C, Potten CS. The intestinal epithelial stem cell. Bioessays 2002; 24(1):91-98.
9. Bach SP, Renehan AG, Potten CS. Stem cells: the intestinal stem cell as a paradigm. Carcinogenesis 2000; 21(3):469-476.
10. Ireland H, Kemp R, Houghton C et al. Inducible Cre-mediated control of gene expression in the murine gastrointestinal tract: effect of loss of beta-catenin. Gastroenterology 2004; 126(5):1236-1246.
11. Andreu P, Colnot S, Godard C et al. Crypt-restricted proliferation and commitment to the paneth cell lineage following apc loss in the mouse intestine. Development 2005; 132(6):1443-1451.
12. Campbell SJ, Carlotti F, Hall PA et al. Regulation of the CYP1A1 promoter in transgenic mice: an exquisitely sensitive on-off system for cell specific gene regulation. J Cell Sci 1996; 109(Pt 11):2619-2625.
13. Sansom OJ, Reed KR, Hayes AJ et al. Loss of Apc in vivo immediately perturbs wnt signaling, differentiation and migration. Genes Dev 2004; 18(12):1385-1390.
14. Sansom OJ, Reed KR, van de Wetering M et al. Cyclin D1 is not an immediate target of beta-catenin following apc loss in the intestine. J Biol Chem 2005; 280(31):28463-28467.
15. Dikovskaya D, Schiffmann D, Newton IP et al. Loss of APC induces polyploidy as a result of a combination of defects in mitosis and apoptosis. J Cell Biol 2007; 176(2):183-195.
16. Bienz M, Clevers H. Linking colorectal cancer to Wnt signaling. Cell 2000; 103(2):311-320.
17. Pinto D, Gregorieff A, Begthel H et al. Canonical wnt signals are essential for homeostasis of the intestinal epithelium. Genes Dev 2003; 17(14):1709-1713.
18. Gregorieff A, Pinto D, Begthel H et al. Expression pattern of wnt signaling components in the adult intestine. Gastroenterology 2005; 129(2):626-638.
19. Reed KR, Sansom OJ, Hayes AJ et al. PPARdelta status and apc-mediated tumourigenesis in the mouse intestine. Oncogene 2004; 23(55):8992-8996.
20. He TC, Chan TA, Vogelstein B et al. PPARdelta is an APC-regulated target of nonsteroidal anti-inflammatory drugs. Cell 1999; 99(3):335-345.
21. Harman FS, Nicol CJ, Marin HE et al. Peroxisome proliferator-activated receptor-delta attenuates colon carcinogenesis. Nat Med 2004; 10(5):481-483.
22. Tetsu O, McCormick F. Beta-catenin regulates expression of cyclin D1 in colon carcinoma cells. Nature 1999; 398(6726):422-426.
23. Labbe E, Lock L, Letamendia A et al. Transcriptional cooperation between the transforming growth factor-beta and wnt pathways in mammary and intestinal tumorigenesis. Cancer Res 2007; 67(1):75-84.
24. Wilson A, Murphy MJ, Oskarsson T et al. c-myc controls the balance between hematopoietic stem cell self-renewal and differentiation. Genes Dev 2004; 18(22):2747-2763.
25. Pelengaris S, Khan M, Evan GI. Suppression of myc-induced apoptosis in beta cells exposes multiple oncogenic properties of myc and triggers carcinogenic progression. Cell 2002; 109(3):321-334.
26. Davis AC, Wims M, Spotts GD et al. A null c-myc mutation causes lethality before 10.5 days of gestation in homozygotes and reduced fertility in heterozygous female mice. Genes Dev 1993; 7(4):671-682.
27. Bettess MD, Dubois N, Murphy MJ et al. c-myc is required for the formation of intestinal crypts but dispensable for homeostasis of the adult intestinal epithelium. Mol Cell Biol 2005; 25(17):7868-7878.
28. Muncan V, Sansom OJ, Tertoolen L et al. Rapid loss of intestinal crypts upon conditional deletion of the wnt/tcf-4 target gene c-myc. Mol Cell Biol 2006; 26(22):8418-8426.
29. Sansom OJ, Meniel VS, Muncan V et al. Myc deletion rescues apc deficiency in the small intestine. Nature 2007; 446(7136):676-679.
30. Batlle E, Henderson JT, Beghtel H et al. Beta-catenin and TCF mediate cell positioning in the intestinal epithelium by controlling the expression of EphB/ephrinB. Cell 2002; 111(2):251-263.

31. Malliri A, Rygiel TP, van der Kammen RA et al. The rac activator tiam1 is a wnt-responsive gene that modifies intestinal tumor development. J Biol Chem 2006; 281(1):543-548.
32. Mogensen MM, Tucker JB, Mackie JB et al. The adenomatous polyposis coli protein unambiguously localizes to microtubule plus ends and is involved in establishing parallel arrays of microtubule bundles in highly polarized epithelial cells. J Cell Biol 2002; 157(6):1041-1048.
33. Nathke I. Cytoskeleton out of the cupboard: colon cancer and cytoskeletal changes induced by loss of APC. Nat Rev Cancer 2006; 6(12):967-974.
34. Kaplan KB, Burds AA, Swedlow JR et al. A role for the adenomatous polyposis coli protein in chromosome segregation. Nat Cell Biol 2001; 3(4):429-432.
35. Smits R, Kielman MF, Breukel C et al. Apc1638T: a mouse model delineating critical domains of the adenomatous polyposis coli protein involved in tumorigenesis and development. Genes Dev 1999; 13(10):1309-1321.
36. Lee HC, Kim M, Wands JR. Wnt/frizzled signaling in hepatocellular carcinoma. Front Biosci 2006; 11:1901-1915.
37. Benhamouche S, Decaens T, Godard C et al. Apc tumor suppressor gene is the "zonation-keeper" of mouse liver. Dev Cell 2006; 10(6):759-770.
38. Tan X, Behari J, Cieply B et al. Conditional deletion of beta-catenin reveals its role in liver growth and regeneration. Gastroenterology 2006; 131(5):1561-1572.
39. Colnot S, Decaens T, Niwa Kawakita M et al. Liver-targeted disruption of apc in mice activates beta-catenin signaling and leads to hepatocellular carcinomas. Proc Natl Acad Sci USA 2004; 101(49):17216-17221.
40. Harada N, Miyoshi H, Murai N et al. Lack of tumorigenesis in the mouse liver after adenovirus-mediated expression of a dominant stable mutant of beta-catenin. Cancer Res 2002; 62(7):1971-1977.
41. Harada N, Oshima H, Katoh M et al. Hepatocarcinogenesis in mice with beta-catenin and ha-ras gene mutations. Cancer Res 2004; 64(1):48-54.
42. Li F, Xiang Y, Potter J et al. Conditional deletion of c-myc does not impair liver regeneration. Cancer Res 2006; 66(11):5608-5612.
43. Pecina Slaus N, Pavelic K, Pavelic J. Loss of heterozygosity and protein expression of APC gene in renal cell carcinomas. J Mol Med 1999; 77(5):446-453.
44. Kim YS, Kang YK, Kim JB et al. beta-catenin expression and mutational analysis in renal cell carcinomas. Pathol Int 2000; 50(9):725-730.
45. Qian CN, Knol J, Igarashi P et al. Cystic renal neoplasia following conditional inactivation of apc in mouse renal tubular epithelium. J Biol Chem 2005; 280(5):3938-3945.
46. Sansom OJ, Griffiths DF, Reed KR et al. Apc deficiency predisposes to renal carcinoma in the mouse. Oncogene 2005; 24(55):8205-8210.
47. Sansom OJ, Meniel V, Wilkins JA et al. Loss of apc allows phenotypic manifestation of the transforming properties of an endogenous K-ras oncogene in vivo. Proc Natl Acad Sci USA 2006; 103(38):14122-14127.
48. Jin Z, Tamura G, Tsuchiya T et al. Adenomatous polyposis coli (APC) gene promoter hypermethylation in primary breast cancers. Br J Cancer 2001; 85(1):69-73.
49. Virmani AK, Rathi A, Sathyanarayana UG et al. Aberrant methylation of the adenomatous polyposis coli (APC) gene promoter 1A in breast and lung carcinomas. Clin Cancer Res 2001; 7(7):1998-2004.
50. Selbert S, Bentley DJ, Melton DW et al. Efficient BLG-cre mediated gene deletion in the mammary gland. Transgenic Res 1998; 7(5):387-396.
51. Gallagher RC, Hay T, Meniel V et al. Inactivation of apc perturbs mammary development, but only directly results in acanthoma in the context of Tcf-1 deficiency. Oncogene 2002; 21(42):6446-6457.
52. Meniel V, Hay T, Douglas Jones A et al. Mutations in apc and p53 synergize to promote mammary neoplasia. Cancer Res 2005; 65(2):410-416.
53. Sansom OJ, Stark LA, Dunlop MG et al. Suppression of intestinal and mammary neoplasia by lifetime administration of aspirin in apc(Min/+) and apc(Min/+), Msh2(−/−) mice. Cancer Res 2001; 61(19):7060-7064.
54. Mahmoud NN, Boolbol SK, Dannenberg AJ et al. The sulfide metabolite of sulindac prevents tumors and restores enterocyte apoptosis in a murine model of familial adenomatous polyposis. Carcinogenesis 1998; 19(1):87-91.
55. Mahmoud NN, Dannenberg AJ, Mestre J et al. Aspirin prevents tumors in a murine model of familial adenomatous polyposis. Surgery 1998; 124(2):225-231.

Chapter 10

Extra-Colonic Manifestations of Familial Adenomatous Polyposis Coli

Alison H. Trainer*

Abstract

The most important clinical manifestation of Familial Adenomatous Polyposis (FAP) is a profuse polyposis predisposing to colorectal cancer. This clinical picture emphasises the fundamental cellular role of APC in colorectal tumorigenesis. As prophylactic colorectal surgery has significantly reduced the mortality associated with this disorder, other less penetrant signs and symptoms of the condition are becoming more clinically relevant. Highlighting these problems will help improve the quality of life for patients with FAP and may have the additional benefit of shedding new light on other possible functions of the APC protein.

Introduction

First described in 1847, familial adenomatous polyposis (FAP) is the archetypal colorectal cancer predisposition syndrome, characterised by the early development of hundreds to thousands of adenoma throughout the large bowel. Left untreated, colorectal cancer usually develops before the age of forty with a severely decreased life expectancy, but usually not before the next generation is affected.

FAP is a highly penetrant, autosomal dominant condition that is due to heterozygous germline mutations in the *adenomatous polyposis coli* (*APC*) gene. Abrogation of the second, wildtype *APC* allele in the colonic epithelium is associated with the formation of numerous premalignant adenomas. Although autosomal dominant in inheritance, a third of affected individuals develop the condition due to a de novo *APC* mutation, with no prior family history or experience of the condition. Understandably, the main clinical focus at diagnosis is on genetic counselling and molecular testing within the families and reducing the colorectal cancer (CRC) risk for the affected individual.

Surgery is the mainstay of management. Often the polyps are too numerous and confluent to be removed individually and the most effective option for the patient is to undergo surgical removal of the colon with preservation of the rectum or, in cases with severe rectal involvement, additional removal of the rectum and the formation of a 'pseudo' rectum/pouch from the small bowel. The timing of the operation is discussed and performed before the risk of developing CRC is high and at a time when the operation and subsequent recuperation period will impact least on the individual's social development, education and career plans.

Prophylactic surgery results in a substantial increase in the life expectancy of patients with FAP; however, this is usually still less than that in the general population. This is due to extracolonic manifestations of FAP[1]; with duodenal carcinoma and desmoids as the most common causes of death after colorectal cancer.[1-3]

*Alison H. Trainer—Prince of Wales Clinical School, Faculty of Medicine. University of New South Wales, Australia. Email: alisonh.trainer@googlemail.com

APC Proteins, edited by Inke S. Näthke and Brooke M. McCartney.

Extended clinical studies have revealed that the clinical spectrum of mutations at the *APC* locus is much wider than first thought[4-6] and that mutations within the *APC* gene may be more common in the population than originally anticipated. In addition, the expanding body of work on APC functions that may affect sites other than the colonic crypt suggests that loss of APC may result in subtle extracolonic phenotypes that our clinical experience to date has not identified. Even as early as 1912, Devic and Bussy described the combination of FAP with extra-intestinal lesions such as soft tissue tumours and multiple benign bone growths (osteomas).[7] This original triad was further extended by Gardner with the recognition of dental anomalies, desmoids and skeletal changes associated with FAP.[8] This combination of polyposis with a high incidence of extra-colonic features was classically known as Gardner syndrome.

The extracolonic manifestations described in FAP patients involve tissue that derives from all three embryonic lineages including the mesoderm: desmoids, osteomas and dental anomalies; endoderm: adenoma and adenocarcinomas of the upper gastrointestinal tract, liver and endocrine tumours; ectoderm: epidermoid cysts of the skin and congenital hypertrophy of the retinal pigment epithelium (CHRPE). Interestingly, phenotype/genotype correlations have been mapped for some clinical presentations as shown in Figure 1. In the next section I will describe some of these extracolonic manifestations in more detail.

Desmoids

Desmoids are slow growing fibrous tumours that, although nonmetastatic, are highly locally invasive, resistant to therapy and prone to recurrence after resection. They occur extremely infrequently in the general population, but arise in about 13% of patients with FAP[9] and are one of the most common causes of mortality and morbidity after colorectal cancer in these patients. They often arise in the abdominal wall and mesentery and slowly infiltrate local organs. Their natural history is variable with periods of minimal growth and some apparent spontaneous regression. They are significantly more common in females, in patients with a family history of desmoids, osteoma or dermoid cysts, and after intra-abdominal surgery.[10] In the classical form of FAP they usually arise singly in the mesentery of the small bowel. There is a strong correlation between desmoids and *APC* genotype: patients with mutations after codon 1444 are 12 times more likely to develop desmoids than individuals with an *APC* mutation more 5′ in the gene[11] (Fig.

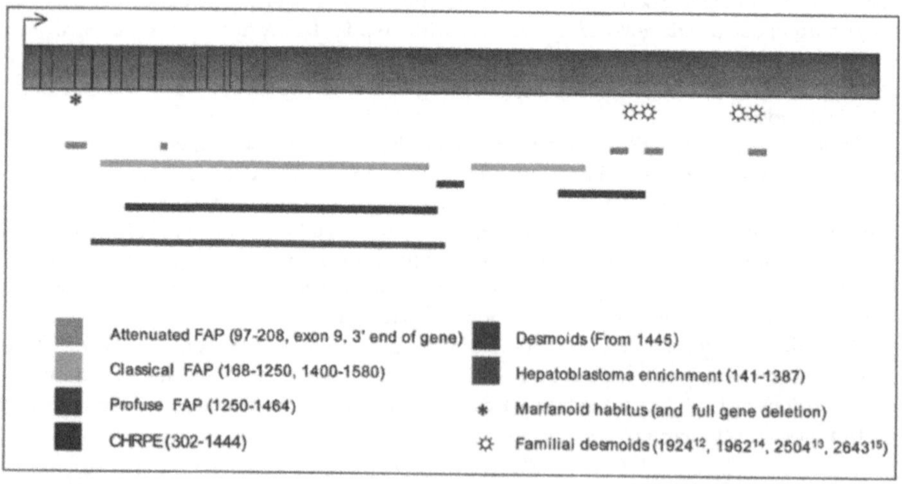

Figure 1. The processed APC transcript comprises 2,843 codons. Shown on the schematic are the positions of published common clinical presentations of *APC* germline mutations along the transcript. Families with identical germline mutations can present with different clinical signs and symptoms, showing that there is significant intra- and inter-familial variation.

1). Consistent with this correlation, families have been documented in whom an autosomal dominant, 100% penetrant form of hereditary desmoids was associated with mutations at the 3′ end of the *APC* gene. In these families, the desmoid formation was more florid and arose near the axial skeleton, rather than intra-abdominally. Only a few polyps were present in specific individuals within the family as the colonic phenotype was uncharacteristically less penetrant than the extracolonic features.[12-15] Although in these desmoid-predisposed families, the germline mutant allele retained at least two of the 20 amino acid repeats which are required for binding β-catenin, some desmoids show a somatic alteration or deletion of the *APC* locus within the tumours[14,15] and show β-catenin staining in both cytoplasm and nucleus.[15] Similar to the evidence in CRC and duodenal tumours, there is an association between the position of the germline mutation within the *APC* locus and the subsequent somatic *APC* mutation within the tumour. Desmoids from patients with FAP consistently retain at least two 20 amino acid β-catenin binding sites or one allele with loss of all 20 amino acid binding sites on the other allele.[16,17] Although this particular correlation between somatic and germline mutant allele is unique to desmoids,[16,17] it follows the same 'somatic mutation preconditioned by germline mutation' model as in the other FAP-associated gastrointestinal tumours, indicating a selective regulation of specific APC functions within these tumours.[17]

Somatic *APC* mutations in conjunction with either LOH or retention of the wildtype allele[18] and indeed activating *β-catenin* mutations,[19] have been described in some sporadic desmoids tumours. The extent to which loss of specific APC-associated pathways contribute to sporadic desmoid formation is uncertain,[20] as immunohistochemistry shows β-catenin in both the cytoplasm and nucleus of the cells, whilst desmoid cells express N-cadherin rather than the E-cadherin expressed in colonic epithelial cells.[18]

Medical treatment regimes used for treatment of desmoids include chemotherapeutic agents such as doxorubicin and dacarbazine,[21] non steroidal anti-inflammatory agents and, based on the higher incidence of desmoids in women, oestrogen modifying agents like Tamoxifen. These treatments are often given in progressive disease to defer surgery, which is associated with a high recurrence rate. These agents can cause regression or stabilisation of the tumour growth.[21-23] However, strong evidence for the clinical utility of these drugs is sparse as the scarcity of the condition means there is no random control trial data available and the variable natural history of the condition confounds interpretation of retrospective studies.

Congenital Hypertrophy of the Retinal Pigment Epithelium

Congenital hypertrophy of the retinal pigment epithelium (CHRPE) was first mentioned in 1975 by Buettner[24] to describe flat pigmented lesions in the retinal pigment epithelium. These lesions are generally benign although they can occasionally cause visual field defects. In the general population, CHRPE manifests as a single, unilateral lesion that is usually detected during routine ocular examination. The reported incidence of single unilateral CHRPE in the general population varies greatly from 0.3 to 40% depending on how actively these lesions are sought. These lesions are not associated with an increase in CRC risk.[25] A subtype of CHRPE was described in a subgroup of families as a congenital, nonprogressive, diagnostic marker for FAP whose presence often predated other clinical manifestations.[26] In contrast to the CHRPE arising in the general population, FAP-associated CHRPEs are often bilateral, multifocal, often with a hypo-pigmented halo with underlying RPE hyperplasia and thickening of Bruch's membrane.[27] Interestingly, in contrast to sporadic CHRPE lesions, in FAP patients the large majority of lesions tend to occur near the retinal vessels.[28] In addition, about 10% of small satellite lesions are only detected by fluorescein angiography.[29] A reclassification of these types of lesions as 'multiple retinal pigment epithelial hamartomata' has been proposed.[30] From a clinical perspective, they form a useful non-invasive diagnostic marker whose presence in a patient predicts a specific subset of *APC* mutations that lie between codons 311 and 1465.[31-33]

Oral and Maxillofacial Manifestations

Although often seen as an isolated anomaly, supernumerary teeth are commonly present in individuals with a diagnosis of cleidocranial dysostosis, syndromic cleft lip and palate and FAP.[34] The frequency of supernumerary teeth in patients with FAP is 11-27%, significantly higher than in the normal population (0-4%).[35,36] In normal tooth development, the proliferating dental lamina extends outgrowths into the ectomesenchyme at sites of odontogenesis. As the tooth develops, the dental lamina and the projecting cells, which connect the developing tooth to the oral epithelium, fragment and undergo apoptosis. In humans, numerous theories exist to explain the pattern and incidence of supernumerary teeth. One proposed mechanism evokes failure of the dental lamina remnants to degenerate and instead to re-activate to form accessory tooth organs. Alternatively, it is possible that the intact dental lamina continues to proliferate due to failure of the epithelial and mesenchymal signalling that normally induces apoptosis. Lastly, the possibility that supernumerary teeth arise from the division of a single tooth bud is supported by a few case reports (review[37]). However, recent mouse work, focussed on addressing these questions, indicates that the Wnt pathway is active at multiple stages in tooth development. Constitutive activation of this pathway, by inhibition of GSK3-β in explant culture or expression of a nondegradable form of β-catenin, is associated with expanded expression of key signalling molecules for normal dentation, Lef1, Bmp4, Msx1 and Msx2 and with abnormal and ectopic tooth initiation, whilst inhibition of the pathway by Dickkopf (*Dkk1*) produced a patterning defect in tooth development that was associated with a reduction in signalling molecules.[38] A similar pattern of tooth developmental arrest associated with *DKK1* expression is also seen in *Msx1–/–*, *Msx2–/–* mice suggesting that this phenotype directly correlates with loss of the Wnt/β-catenin/Lef1 transcription activation pathway.[39]

This work underlines the importance of both the Wnt pathway itself as well as its exquisite regulation in tooth development. Interestingly, the phenotype in the patient with FAP is assumed to arise in the context of cells with a heterozygous *APC* mutation. Heterozygous FAP mouse models show no abnormality in tooth development whilst the *APC* homozygote dental phenotype has not been examined (Owen Sampson, personal communication).

Odontomas, well-defined encapsulated hard tissue growths with an odontogenic appearance, are also more frequently present in the mandible and in the maxilla of FAP patients than the normal population.[35,40] The highest incidence of supernumerary teeth and odontomas is seen in FAP patients with more than three osteomas,[35] which is consistent with the finding that patients with widespread osteomatous lesions have a higher frequency of odontomas than patients with a single localised lesion.[40] Although, in patients with mutations 3′ to codon 1444 more dental anomalies are detected on dental panoramic radiographs than in patients with mutations more 5′ in the gene,[41] a correlation between dental abnormalities and colonic adenomas has not been detected so far.[42] The APC mouse models show no dental abnormalities, but as murine dentition is severely reduced compared to human, lacking premolars and canines, the validity of the mouse as suitable model in this case is not clear.

Osteomas are detected in 46-93% of patients with FAP; a prevalence that is 4-20X higher than in the normal population.[40,43-45] They commonly occur in both the upper and lower jaws, although they seem to have a predilection for the mandible and vary in size from 3 mm to 4 cm.[46,47] They occur around puberty and can show a slow growth pattern after adolescence.[35,45] Most patients have osteomas enclosed within the normal bone, i.e., enostoses and these are commonly found in the premolar-molar region.[40,46] However, in a third of patients these lesions protrude, i.e., exostosis, around the mandibular angle or the zygomatic arches.[35] Most osteomas are asymptomatic, although their growth may impinge on local function such as oral movement and produce palatal and speech defects. Restriction of movement and cosmetic disfigurement are the two most common reasons for surgery.

Extra-Gastrointestinal Tumour Predisposition

The three main sites for extra-colonic tumours in patients with germline mutations in the *APC* gene locus are brain, thyroid and liver. Although the relative risk of developing these cancers is greatly increased in patients with FAP compared to the general population, their absolute incidence within the FAP patient population is still low. Consistent with this low penetrance, these tumours themselves seldom show somatic loss or mutation of the wildtype *APC* gene locus, suggesting an interaction between *APC* haploinsufficiency and either an environmental or modifier gene effect on tumour initiation/evolution.

Turcot syndrome (brain tumour-polyposis syndrome) is characterised by the concurrence of primary brain tumours and colorectal adenomas. One third of the families that present with this disorder are diagnosed with hereditary nonpolyposis coli (HNPCC) syndrome, which is due to an aberrant mismatch repair pathway, whilst two thirds are due to FAP. The main CNS tumours associated with FAP are medulloblastoma, astrocytoma and, less frequently, ependymona and pinealoblastoma.[48] The incidence of CNS tumours is low, although the relative risk for a CNS tumour and more specifically a medulloblastoma in patients with FAP is increased 7 and 92 times,[49] respectively compared to that of the general population. Despite the accepted clinical association between FAP and medulloblastomas, there is little conclusive evidence of LOH or mutation of the somatic wildtype *APC* allele in FAP-associated CNS tumours.[49,50] In addition, neither somatic *APC* mutations[50-52] nor *axin1* mutations [53,54] are common in sporadic medulloblastoma. Similarly, despite aberrant β-catenin staining in 25% of medulloblastoma, less than 10% of these tumours have an oncogenic mutation in β-catenin itself.[51,55] Although a correlation between *APC* mutations between codons 679-1224 and an increased risk of CNS tumours has been proposed recently,[48] the molecular aetiology of medulloblastoma in the context of FAP patients is still unclear. Specifically how aberrant Wnt signalling or the loss of other APC functions contribute to these tumours is not known.

Hepatoblastoma is a rapidly progressive malignant embryonic liver tumour that classically occurs in the first few years of life. It has a reasonable prognosis with overall survival rate of 65-78%. Although the overall incidence is only 0.82% to 1.5% in FAP patients, with a male predominance,[56,57] this represents an increased relative risk of 750-7,500 times compared to the normal population. Although the site of the *APC* mutation does not predict the occurrence of this tumour within FAP families, the germline mutations associated with hepatoblastoma are enriched between codon 141 to 1387.[56-58] Contrary to hepatocellular carcinoma, hepatoblastomas are often diploid or hyperdiploid and harbour only a limited number of chromosomal abnormalities. More than half of these tumours express a mutant form of β-catenin with interstitial deletions or mis-sense mutations within the GSK3β phosphorylation site. This mutant protein was found to be nuclear on immunohistochemistry[59](review 60). There is little information about the somatic *APC* allele in FAP associated with hepatoblastomas. Although somatic mutations of the wildtype allele have been reported,[61] it is not always present.[62]

Consistent with activation of the Wnt signalling pathway, increased expression of cyclin D1 has been demonstrated in primary hepatoblastomas, although interestingly another known β-catenin regulated gene, *c-myc*, was not.[59] Somatic mutations within the *APC* gene locus have been found in a limited number of sporadic hepatoblastomas only.[63]

Thyroid cancer occurs in approximately 1% of FAP patients with a high female predominance and is often associated with extra-colonic features.[64] Characteristically, it is multi-centric, occurs earlier than its sporadic counterpart and is often associated with an unusual papillary cribriform variant.[65] The tumours rarely show LOH of the wildtype *APC* allele, although missense somatic mutations have been described.[66,67] There is a high rate of rearrangement within the *RET* proto-oncogene found within the tumours, which results in constitutive expression of the Ret oncogene.[65,68,69] Rearrangements in the *RET* locus is found in 25-40% of papillary thyroid carcinoma.[70] This suggests that the predisposition to papillary thyroid carcinoma in FAP patients, unlike that of colorectal cancer, is associated with *APC* haplo-insufficiency. This haplo-insufficiency may cooperate with constitutive RET activation. However, it should be noted that in addition FAP-associated and sporadic papillary cribriform thyroid cancer are associated with somatic mutations within exon 3 of the *β-catenin* locus and the accumulation of mutant β-catenin.[67]

Marfanoid Habitus

Intriguingly, three case reports describe individuals with profuse early onset polyposis and clinical features suggestive of Marfan syndrome. Marfan syndrome (OMIM 154700) is an inherited connective tissue disorder due to germline mutations in the *Fibrillin-1*(*FBN1*) gene locus. The clinical severity of the condition is highly variable but the cardinal manifestations involve the ocular, skeletal and cardiovascular systems. The skeletal system changes are characterized by excessive linear growth of the long bones and joint laxity. The extremities are disproportionately long for the size of the trunk leading to a reduction in the arm span-to-height and upper-to-lower segment ratios. Overgrowth of the ribs can also alter the shape of the sternum. Patients may also present with dislocated lenses. However, the major cause of morbidity and mortality or Marfan syndrome is associated with cardiovascular complications. These comprise dilation of the aorta, which can lead to rupture, as well as a tendency to other cardiac valve prolapses.

The association of polyposis and Marfanoid features in these three patients is suggestive of an underlying connective tissue disorder. Two of the three cases had a documented *APC* mutation. Case one had a deletion of the *APC* gene locus associated with Marfan-like skeletal disproportions, an abnormal ribcage, wasted calf muscles and Caroli syndrome in addition to a florid polyposis. The *APC* deletion did not involve the adjacent *Fibrillin 2* (*FBN2*) gene locus; excluding the possibility that deletion of this locus was the cause of the marfanoid features.[71] The second case, a 38 year old man, whose mother also had a marfanoid habitus and had died from gastrointestinal malignancy at the age of 34, was found to have a marfanoid skeletal disproportion with the cardiovascular complications of a dilated aortic root, mitral valve prolapse and an atrial septal aneurysm. In addition, he had supernumerary teeth and a mesenchymal hamartoma. His affected sister was described as tall and thin but with no other clinical stigmata of Marfan syndrome. A donor splice site mutation in exon 4 of *APC* was described in this family. RNA studies from this family confirmed a transcript lacking exon 4 resulting in a frame shift. However, no truncated protein was detected in immunoblots.[72] The last case also presented with features of Marfan syndrome, high arch palate, thoracic kyphoscoliosis, moderate hypermobility and a cardiovascular murmur but no DNA was available for mutation analysis.

The role of APC in the regulation of nuclear β-catenin levels and thus in Wnt signalling is known to be important in colonic polyp formation. Thus, most extracolonic manifestation have been examined with regard to either loss or retention of the ability of the mutant APC to bind β-catenin, or the cellular localisation of β-catenin by immunohistochemistry, which is often not informative. Assays for other APC functions, microtubule binding etc. are less amenable to routine functional scrutiny but it is noteworthy that inherent in truncations deleting even one β-catenin binding site in the APC protein, is the loss of all carboxyl terminus protein interactions.

The decreased penetrance of many extracolonic features suggests a less direct correlation with APC function in these target cells or cell type specific effects. The identification of β-catenin activating mutations in sporadic hepatoblastomas does suggest that Wnt signalling is important in the formation of these tumours, whilst the absence of somatic *APC* or β-catenin mutations in the majority of thyroid cancers suggests that the development of thyroid cancer in FAP patients is not dependent on this pathway.

A complicating factor is the possibility that genetic modifier genes may account for the family-specific nature of many extracolonic features. This is particularly relevant in the case of desmoid formation.

Now that individuals with familial adenomatous polyposis live considerably longer due to the success of preventative colorectal surgery, a new clinical approach to these patients is merited. A finer focus on the more subtle clinical manifestations of FAP is required. By concentrating on the more covert as well as overt extra-intestinal manifestations of the disease, we should be able to improve the quality as well quantity of life for patients with familial adenomatous polyposis coli.

References

1. Nugent KP, Spigelman AD, Phillips RK. Life expectancy after colectomy and ileorectal anastomosis for familial adenomatous polyposis. Dis Colon Rectum 1993; 36:1059-62.
2. Arvanitis ML, Jagelman DG, Fazio VW et al. Mortality in patients with familial adenomatous polyposis. Dis Colon Rectum 1990; 33:639-42.
3. Bertario L, Presciuttini S, Sala P et al. Causes of death and postsurgical survival in familial adenomatous polyposis: results from the italian registry. Italian registry of familial polyposis writing committee. Semin Surg Oncol 1994; 10:225-34.
4. Spirio L, Calin G, Olschwang S et al. Alleles of the APC gene: an attenuated form of familial polyposis. Cell 1993; 75:951-7.
5. Soravia C, Berk T, Madlensky L et al. Genotype-phenotype correlations in attenuated adenomatous polyposis coli. Am J Hum Genet 1998; 62:1290-301.
6. Moisio AL, Jarvinen H, Peltomaki P. Genetic and clinical characterisation of familial adenomatous polyposis: a population based study. Gut 2002; 50:845-50.
7. Devic A, Bussy MM. Un cas de polypose adenomateuze generalisee autour de l'intestine. Arch Mal Appar Dig 1912; 6:278-299.
8. Gardner EJ, Richards RC. Multiple cutaneous and subcutaneous lesions occurring simultaneously with hereditary polyposis and osteomatosis. Am J Hum Genet 1953; 5:139-47.
9. Klemmer S, Pascoe L, DeCosse J. Occurrence of desmoids in patients with familial adenomatous polyposis of the colon. Am J Med Genet 1987; 28:385-92.
10. Sagar PM, Moslein G, Dozois RR. Management of desmoid tumors in patients after ileal pouch-anal anastomosis for familial adenomatous polyposis. Dis Colon Rectum 1998; 41:1350-5; discussion 1355-6.
11. Bertario L, Russo A, Sala P et al. Genotype and phenotype factors as determinants of desmoid tumors in patients with familial adenomatous polyposis. Int J Cancer 2001; 95:102-7.
12. Scott RJ, Froggatt NJ, Trembath RC et al. Familial infiltrative fibromatosis (desmoid tumours) (MIM135290) caused by a recurrent 3' APC gene mutation. Hum Mol Genet 1996; 5:1921-4.
13. Eccles D, Harvey J, Bateman A et al. A novel 3' mutation in the APC gene in a family presenting with a desmoid tumour. J Med Genet 2001; 38:861-3.
14. Eccles DM, Van Der Luijt R, Breukel C et al. Hereditary desmoid disease due to a frameshift mutation at codon 1924 of the APC gene. Am J Hum Genet 1996; 59:1193-201.
15. Couture J, Mitri A, Lagace R et al. A germline mutation at the extreme 3' end of the APC gene results in a severe desmoid phenotype and is associated with overexpression of beta-catenin in the desmoid tumor. Clin Genet 2000; 57:205-12.
16. Latchford A, Volikos E, Johnson V et al. APC mutations in FAP-associated desmoid tumours are nonrandom but not 'just right'. Hum Mol Genet 2007; 16:78-82.
17. Miyaki M, Yamaguchi T, Iijima T et al. Difference in characteristics of APC mutations between colonic and extracolonic tumors of FAP patients: variations with phenotype. Int J Cancer 2008; 122:2491-7.
18. Alman BA, Li C, Pajerski ME et al. Increased beta-catenin protein and somatic APC mutations in sporadic aggressive fibromatoses (desmoid tumors). Am J Pathol 1997; 151:329-34.
19. Shitoh K, Konishi F, Iijima T et al. A novel case of a sporadic desmoid tumour with mutation of the beta catenin gene. J Clin Pathol 1999; 52:695-6.
20. Giarola M, Wells D, Mondini P et al. Mutations of adenomatous polyposis coli (APC) gene are uncommon in sporadic desmoid tumours. Br J Cancer 1998; 78:582-7.
21. Gega M, Yanagi H, Yoshikawa R et al. Successful chemotherapeutic modality of doxorubicin plus dacarbazine for the treatment of desmoid tumors in association with familial adenomatous polyposis. J Clin Oncol 2006; 24:102-5.
22. Izes JK, Zinman LN, Larsen CR. Regression of large pelvic desmoid tumor by tamoxifen and sulindac. Urology 1996; 47:756-9.
23. Maroy B. Desmoid tumor sensitive to tamoxifen. Presse Med 1997; 26, 1520-2.
24. Buettner H. Congenital hypertrophy of the retinal pigment epithelium. Am J Ophthalmol 1975; 79:177-89.
25. Shields JA, Shields CL, Shah PG et al. Lack of association among typical congenital hypertrophy of the retinal pigment epithelium, adenomatous polyposis and gardner syndrome. Ophthalmology 1992; 99:1709-13.
26. Blair NP, Trempe CL. Hypertrophy of the retinal pigment epithelium associated with Gardner's syndrome. Am J Ophthalmol 1980; 90:661-7.
27. Tourino R, Conde-Freire R, Cabezas-Agricola JM et al. Value of the congenital hypertrophy of the retinal pigment epithelium in the diagnosis of familial adenomatous polyposis. Int Ophthalmol 2004; 25:101-12.

28. Schmidt D, Jung CE, Wolff G. Changes in the retinal pigment epithelium close to retinal vessels in familial adenomatous polyposis. Graefes Arch Clin Exp Ophthalmol 1994; 232:96-102.
29. Tourino R, Rodriguez-Ares MT, Lopez-Valladares MJ et al. Fluorescein angiographic features of the congenital hypertrophy of the retinal pigment epithelium in the familial adenomatous polyposis. Int Ophthalmol 2005; 26:59-65.
30. Hennessy MP, Collins F, Coroneo MT. The distinction between multiple retinal pigment epithelial hamartomata (MRPEH) in familial adenomatous polyposis (FAP) and congenital hypertrophy of the retinal pigment epithelium (CHRPE). Aust N Z J Ophthalmol 1993; 21:275-6.
31. Enomoto M, Konishi M, Iwama T et al. The relationship between frequencies of extracolonic manifestations and the position of APC germline mutation in patients with familial adenomatous polyposis. Jpn J Clin Oncol 2000; 30:82-8.
32. Giardiello FM, Petersen GM, Piantadosi S et al. APC gene mutations and extraintestinal phenotype of familial adenomatous polyposis. Gut 1997; 40:521-5.
33. Ficari F, Cama A, Valanzano R et al. APC gene mutations and colorectal adenomatosis in familial adenomatous polyposis. Br J Cancer 2000; 82:348-53.
34. Fader M, Kline SN, Spatz SS et al. Gardner's syndrome (intestinal polyposis, osteomas, sebaceous cysts) and a new dental discovery. Oral Surg Oral Med Oral Pathol 1962; 15:153-72.
35. Wolf J, Jarvinen HJ, Hietanen J. Gardner's dento-maxillary stigmas in patients with familial adenomatosis coli. Br J Oral Maxillofac Surg 1986; 24:410-6.
36. Sondergaard JO, Bulow S, Jarvinen H et al. Dental anomalies in familial adenomatous polyposis coli. Acta Odontol Scand 1987; 45:61-3.
37. D'Souza RN, Klein OD. Unraveling the molecular mechanisms that lead to supernumerary teeth in mice and men: current concepts and novel approaches. Cells Tissues Organs 2007; 186:60-9.
38. Liu F, Chu EY, Watt B et al. Wnt/beta-catenin signaling directs multiple stages of tooth morphogenesis. Dev Biol 2008; 313:210-24.
39. Bei M, Maas R. FGFs and BMP4 induce both Msx1-independent and Msx1-dependent signaling pathways in early tooth development. Development 1998; 125:4325-33.
40. Kubo K, Miyatani H, Takenoshita Y et al. Widespread radiopacity of jaw bones in familial adenomatosis coli. J Craniomaxillofac Surg 1989; 17:350-3.
41. Davies DR, Armstrong JG, Thakker N et al. Severe Gardner syndrome in families with mutations restricted to a specific region of the APC gene. Am J Hum Genet 1995; 57:1151-8.
42. Takeuchi T, Takenoshita Y, Kubo K et al. Natural course of jaw lesions in patients with familial adenomatosis coli (Gardner's syndrome). Int J Oral Maxillofac Surg 1993; 22:226-30.
43. Utsunomiya J, Nakamura T. The occult osteomatous changes in the mandible in patients with familial polyposis coli. Br J Surg 1975; 62:45-51.
44. Thakker N, Davies R, Horner K et al. The dental phenotype in familial adenomatous polyposis: diagnostic application of a weighted scoring system for changes on dental panoramic radiographs. J Med Genet 1995; 32:458-64.
45. Ullbro C, Alm T, Ericson S et al. Occult radiopaque jaw lesions in familial adenomatous polyposis. Swed Dent J 1990; 14:201-12.
46. Yuasa K, Yonetsu K, Kanda S et al. Computed tomography of the jaws in familial adenomatosis coli. Oral Surg Oral Med Oral Pathol 1993; 76:251-5.
47. Halling F, Merten HA, Lepsien G et al. Clinical and radiological findings in Gardner's syndrome: a case report and follow-up study. Dentomaxillofac Radiol 1992; 21:93-8.
48. Attard TM, Giglio P, Koppula S et al. Brain tumors in individuals with familial adenomatous polyposis: a cancer registry experience and pooled case report analysis. Cancer 2007; 109:761-6.
49. Hamilton SR, Liu B, Parsons RE et al. The molecular basis of Turcot's syndrome. N Engl J Med 1995; 332:839-47.
50. Mori T, Nagase H, Horii A et al. Germ-line and somatic mutations of the APC gene in patients with turcot syndrome and analysis of APC mutations in brain tumors. Genes Chromosomes Cancer 1994; 9:168-72.
51. Huang H, Mahler-Araujo BM, Sankila A et al. APC mutations in sporadic medulloblastomas. Am J Pathol 2000; 156:433-7.
52. Yong WH, Raffel C, von Deimling A et al. The APC gene in Turcot's syndrome. N Engl J Med 1995; 333:524.
53. Dahmen RP, Koch A, Denkhaus D et al. Deletions of AXIN1, a component of the WNT/wingless pathway, in sporadic medulloblastomas. Cancer Res 2001; 61:7039-43.
54. Baeza N, Masuoka J, Kleihues P et al. AXIN1 mutations but not deletions in cerebellar medulloblastomas. Oncogene 2003; 22:632-6.
55. Zurawel RH, Chiappa SA, Allen C et al. Sporadic medulloblastomas contain oncogenic beta-catenin mutations. Cancer Res 1998; 58:896-9.

56. Giardiello FM, Petersen GM, Brensinger JD et al. Hepatoblastoma and APC gene mutation in familial adenomatous polyposis. Gut 1996; 39:867-9.
57. Hirschman BA, Pollock BH, Tomlinson GE. The spectrum of APC mutations in children with hepatoblastoma from familial adenomatous polyposis kindreds. J Pediatr 2005; 147:263-6.
58. Cetta F, Montalto G, Petracci M. Hepatoblastoma and APC gene mutation in familial adenomatous polyposis. Gut 1997; 41:417.
59. Takayasu H, Horie H, Hiyama E et al. Frequent deletions and mutations of the beta-catenin gene are associated with overexpression of cyclin D1 and fibronectin and poorly differentiated histology in childhood hepatoblastoma. Clin Cancer Res 2001; 7:901-8.
60. Buendia MA. Genetic alterations in hepatoblastoma and hepatocellular carcinoma: common and distinctive aspects. Med Pediatr Oncol 2002; 39:530-5.
61. Kurahashi H, Takami K, Oue T et al. Biallelic inactivation of the APC gene in hepatoblastoma. Cancer Res 1995; 55:5007-11.
62. Inukai T, Furuuchi K, Sugita K et al. Nuclear accumulation of beta-catenin without an additional somatic mutation in coding region of the APC gene in hepatoblastoma from a familial adenomatous polyposis patient. Oncol Rep 2004; 11:121-6.
63. Oda H, Imai Y, Nakatsuru Y et al. Somatic mutations of the APC gene in sporadic hepatoblastomas. Cancer Res 1996; 56:3320-3.
64. Truta B, Allen BA, Conrad PG et al. Genotype and phenotype of patients with both familial adenomatous polyposis and thyroid carcinoma. Fam Cancer 2003; 2:95-9.
65. Soravia C, Sugg SL, Berk T et al. Familial adenomatous polyposis-associated thyroid cancer: a clinical, pathological and molecular genetics study. Am J Pathol 1999; 154:127-35.
66. Cetta F, Curia MC, Montalto G et al. Thyroid carcinoma usually occurs in patients with familial adenomatous polyposis in the absence of biallelic inactivation of the adenomatous polyposis coli gene. J Clin Endocrinol Metab 2001; 86:427-32.
67. Xu B, Yoshimoto K, Miyauchi A et al. Cribriform-morular variant of papillary thyroid carcinoma: a pathological and molecular genetic study with evidence of frequent somatic mutations in exon 3 of the beta-catenin gene. J Pathol 2003; 199:58-67.
68. Cetta F, Chiappetta G, Melillo RM et al. The ret/ptc1 oncogene is activated in familial adenomatous polyposis-associated thyroid papillary carcinomas. J Clin Endocrinol Metab 1998; 83:1003-6.
69. Cetta F, Toti P, Petracci M et al. Thyroid carcinoma associated with familial adenomatous polyposis. Histopathology 1997; 31:231-6.
70. Tallini G, Asa SL. RET oncogene activation in papillary thyroid carcinoma. Adv Anat Pathol 2001; 8:345-54.
71. Hodgson SV, Coonar AS, Hanson PJ et al. Two cases of 5q deletions in patients with familial adenomatous polyposis: possible link with Caroli's disease. J Med Genet 1993; 30:369-75.
72. Calin G, Wijnen J, Van Der Klift H et al. Marfan-like habitus and familial adenomatous polyposis in two unrelated males: a significant association? Eur J Hum Genet 1999; 7:609-14.

Index

GPSR Compliance

The European Union's (EU) General Product Safety Regulation (GPSR) is a set of rules that requires consumer products to be safe and our obligations to ensure this.

If you have any concerns about our products, you can contact us on ProductSafety@springernature.com

In case Publisher is established outside the EU, the EU authorized representative is:

Springer Nature Customer Service Center GmbH
Europaplatz 3
69115 Heidelberg, Germany

Batch number: 10620141

Printed by Printforce, the Netherlands